IMMUNOBIOLOGY OF PROTEINS AND PEPTIDES—II

ADVANCES IN EXPERIMENTAL MEDICINE AND BIOLOGY

Recent Volumes in this Series

Volume 147
INTRAOVARIAN CONTROL MECHANISMS
Edited by Cornelia P. Channing and Sheldon J. Segal

Volume 148
STRUCTURE AND FUNCTION RELATIONSHIPS IN BIOCHEMICAL SYSTEMS
Edited by Francesco Bossa, Emilia Chiancone, Alessandro Finazzi-Agrò, and Roberto Strom

Volume 149
IN VIVO IMMUNOLOGY: Histophysiology of the Lymphoid System
Edited by Paul Nieuwenhuis, A. A. van den Broek, and M. G. Hanna, Jr.

Volume 150
IMMUNOBIOLOGY OF PROTEINS AND PEPTIDES—II
Edited by M. Z. Atassi

Volume 151
REGULATION OF PHOSPHATE AND MINERAL METABOLISM
Edited by Shaul G. Massry, Joseph M. Letteri, and Eberhard Ritz

Volume 152
NEW VISTAS IN GLYCOLIPID RESEARCH
Edited by Akira Makita, Shizuo Handa, Tamotsu Taketomi, and Yoshitaka Nagai

Volume 153
UREA CYCLE DISEASES
Edited by A. Lowenthal, A. Mori, and B. Marescau

IMMUNOBIOLOGY OF PROTEINS AND PEPTIDES—II

Edited by
M. Z. Atassi
Mayo Clinic
Rochester, Minnesota

PLENUM PRESS • NEW YORK AND LONDON

Library of Congress Cataloging in Publication Data

International Symposium on Immunobiology of Proteins and Peptides (2nd: 1981: Tahoe City, Calif.)
Immunobiology of proteins and peptides–II.

(Advances in experimental medicine and biology; v. 150)
"Proceedings of the Second International Symposium on the Immunobiology of Proteins and Peptides, held September 8-11, 1981, in Tahoe City, California" –T.p. verso.
Includes bibliographical references and index.
1. Immune response–Regulation–Congresses. 2. Proteins–Congresses. 3. Peptides–Congresses. I. Atassi, M. Z. II. Title. III. Series. [DNLM: 1. Peptides–Immunology–Congresses. 2. Proteins–Immunology–Congresses. W1 AD559/QW 570 I6733 1981i]
QR186.I56 1981 599.02'9 92-12360
ISBN-13: 978-1-4684-4333-2 e-ISBN-13: 978-1-4684-4331-8
DOI: 10.1007/978-1-4684-4331-8

Proceedings of the Second International Symposium on the Immunobiology of Protein and Peptides, held September 8 – 11, 1981, in Tahoe City, California

Scientific Council of the Symposium

M. Z. Atassi, Chairman
E. Benjamini
C. S. David
K. J. Dorrington
J. E. Goodman
E. Margoliash
A. S. Rosenthal
A. Sehon
E. E. Sercarz
T. B. Tomasi
C. R. Young, Secretary

Symposium sponsors include

Accurate Chemical and Scientific Corporation
Bethesda Research Laboratories, Inc.
E. I. DuPont De Nemours and Company
Genentech, Inc.
Merck, Sharp and Dohm Research Laboratories
Minnesota Mining and Manufacturing Company
Pierce Chemical Company
Waters Associates

Softcover reprint of the hardcover 1st edition 1982
A Division of Plenum Publishing Corporation
233 Spring Street, New York, N.Y. 10013

PREFACE

The immune response is largely dependent on molecular interactions involving proteins. The recognition of antigen molecules, whether they are proteins or non-proteins, whether they are self or non-self, takes place at the molecular-cellular interface through membrane receptor molecules that are proteins. The initial step of recognition activates a complex series of cellular events requiring some mechanism of cell-cell interactions and communications, eventually leading to antibody production. This biological cascade is controlled at several positions along its consecutive pathways by protein molecules, either in the free form or as receptors on membranes of cells committed to this activity. Clearly, then, the proper understanding of the response by cells of the immune system will depend, to a great measure, on the definition of the molecular events involving protein interactions. Obviously, cells work via molecules and molecules work via cells and, at this level of functional resolution, molecular immunology and cellular immunology will merge and will depend heavily on protein chemistry.

These considerations provided the rationale for the foundation of the International Symposium on the Immunobiology of Proteins and Peptides. The word "peptides" has been taken out now from the name simply for the sake of brevity and not to imply any change in emphasis. Proteins in immunology have been handled in various immunological symposia and congresses, frequently scattered as a supporting cast but never the main actors. These symposia series were therefore established to focus on the role of proteins in immunology not only as antigens or immunogens but also as regulators, mediators and end products of the response. By bringing together investigators in molecular and cellular immunology it is hoped that this interaction will help dissolve the communication barrier that has often seemed to separate them. The meeting will focus on current findings and concepts in contemporary immunology and will identify some important new directions for future investigations.

M. Z. Atassi
E. Benjamini

CONTENTS

MOLECULAR ASPECTS OF T- AND B-CELL RECOGNITION OF PROTEIN ANTIGENS

1

AUTOIMMUNITY TO MYELIN BASIC PROTEIN

Robert H. Swanborg, James H. Holda and Joyce A. Killen

Department of Immunology and Microbiology
Wayne State University School of Medicine
Detroit, Michigan 48201

INTRODUCTION

Experimental allergic encephalomyelitis (EAE) is an autoimmune central nervous system (CNS) disease which has been studied extensively, and has provided insight into mechanisms of immune tissue damage and immunologic self-tolerance. In addition, EAE serves as a prototype for human demyelinating diseases such as multiple sclerosis[1].

Autoimmune diseases are thought to arise when self-tolerance mechanisms fail[2]. This permits the break-through of autoreactive lymphocytes which initiate immunologically-mediated damage to the relevant target organ. In EAE, the autoreactive lymphocytes respond against myelin basic protein (BP), a major structural constituent of the myelin sheath, resulting in mononuclear cell inflammation and demyelination in the brain and spinal cord, which are manifested clinically as paralysis. BP can be extracted from myelin and purified by chromatography (Fig. 1; ref. 3). The BP molecule consists of a single polypeptide chain of about 170 amino acids, with a molecular weight of approximately 18,400 daltons. Several investigators have carried out amino acid sequence analysis[4,5], which revealed that the structure of the molecule is highly conserved in mammals. One exception is that two molecular species of BP exist in rodents; the 170 residue (large) BP, and a smaller molecule from which 40 residues have been deleted (small BP)[6].

INDUCTION OF EAE

Fortunately, the occurrence of autoimmune diseases is rare since several self-tolerance mechanisms exist to prevent the prolif-

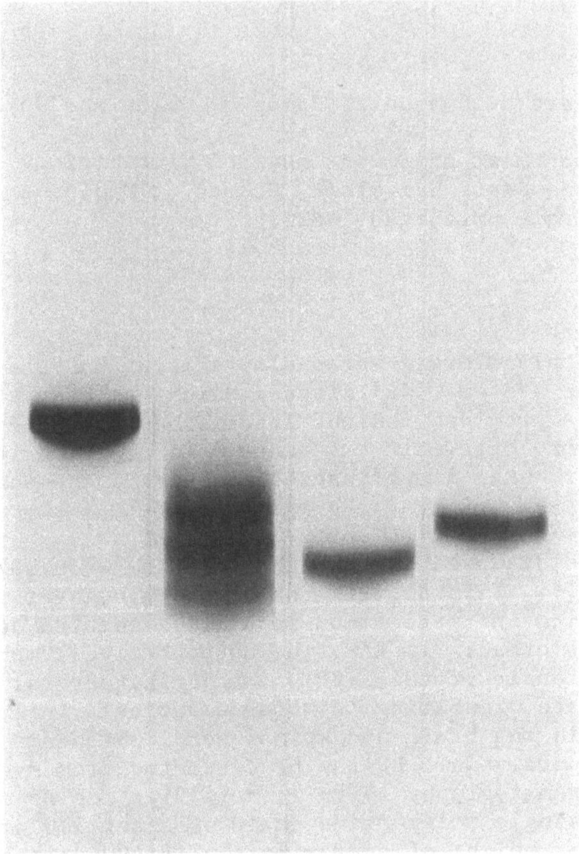

Fig. 1. Polyacrylamide gel electrophoresis. (left-to-right) purified BP; pepsin digest of BP; peptide 44-89; peptide 38-89. The cathode is at the bottom.

eration of autoreactive cells. In EAE, self-tolerance is abrogated by injecting BP with complete Freund's adjuvant (CFA). Although the action of CFA has not been completely elucidated, this adjuvant activates T lymphocytes (including autoreactive T cells), and macrophages which are involved in the effector phase of the autoimmune response.

EAE in the Guinea Pig

Hartley and strain 13 guinea pigs are very susceptible to EAE, whereas strain 2 animals are resistant to the disease. The genetic basis of this susceptibility appears to be centered upon the ability of the animal to respond to BP, and probably resides in the T cells.

One injection of 10-25 μg BP in CFA is sufficient to cause EAE in guinea pigs. Approximately 10 days after sensitization, the animal begins to lose weight (often more than 20 percent of the body weight is lost within 2-3 days), and hind limb paralysis occurs 1 to 3 days later. Paralyzed guinea pigs quickly succumb to the disease. Histologically, one finds extensive mononuclear cell infiltration in the brain and spinal cord.

The major disease-inducing (encephalitogenic) determinant of the BP molecule for guinea pigs has been identified as comprising residues 114-to-122 (sequence: Phe-Ser-Trp-Gly-Ala-Glu-Gly-Gln-Lys), which contains the sole tryptophan residue in the molecule. Accordingly, BP can be rendered non-encephalitogenic for guinea pigs by blocking this tryptophan residue (e.g., by alkylation with 2-hydroxy-5-nitrobenzyl bromide, HNB)[7,8]. Moreover, synthetic peptides corresponding to BP residues 114-122 have been shown to have encephalitogenic activity, although in our hands, these peptides are less encephalitogenic than the intact BP molecule.

Several studies have shown a correlation between cellular immunity to BP and EAE in the guinea pig[9,10]. In contrast, no convincing evidence exists to implicate antibody in the pathogenesis of this disease. Cellular immunity has been demonstrated by skin test reactions to BP (which are maximal about 20-24 hr after intracutaneous injection of antigen), and *in vitro*, by assay for macrophage migration inhibition factor (MIF).

EAE in the Lewis Rat

Like Hartley and strain 13 guinea pigs, Lewis rats are very susceptible to EAE, although the disease differs in at least two respects. Notably, Lewis rats spontaneously recover from EAE and are subsequently resistant to further attempts to reinduce the disease. Moreover, unlike guinea pigs in which EAE can be induced with most mammalian BP's, rats respond best to BP derived from

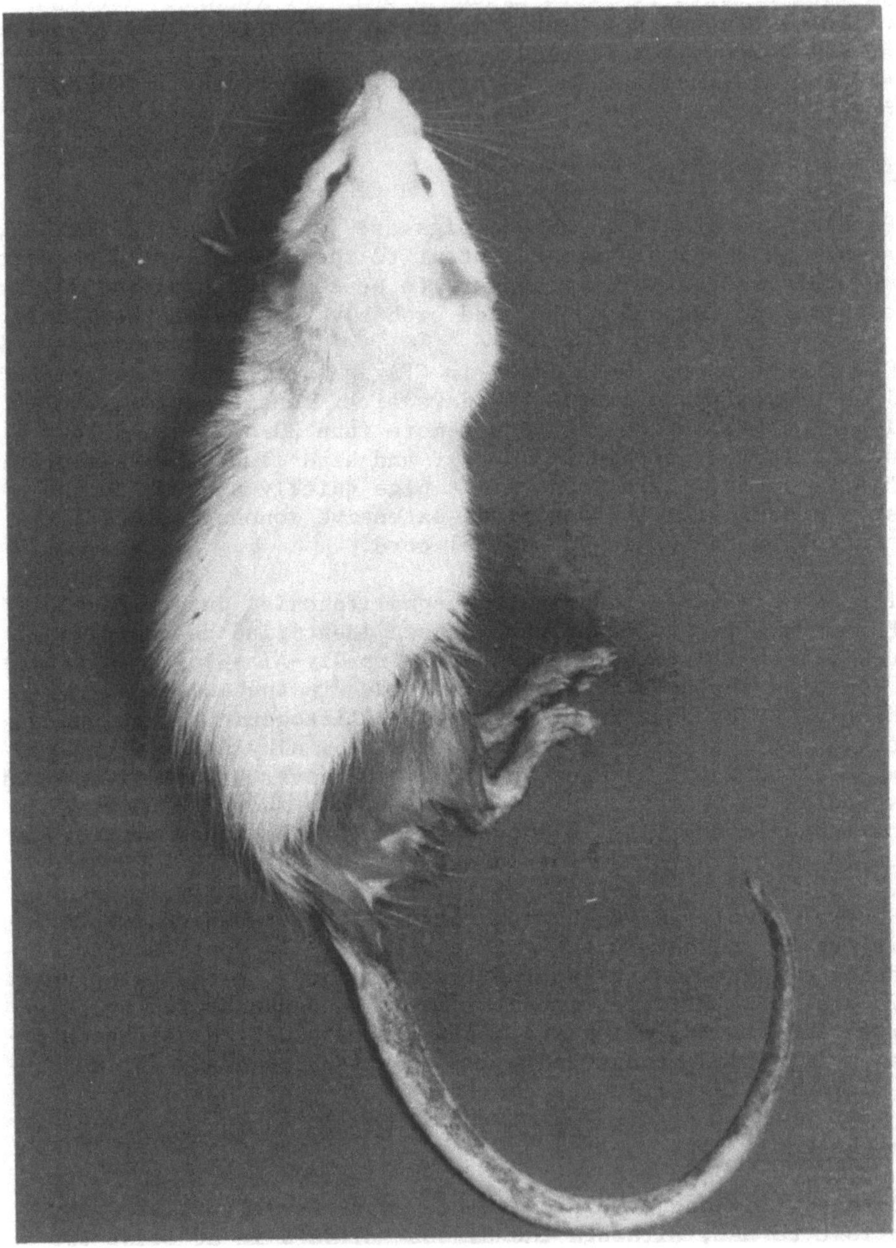

Fig. 2. Lewis rat with EAE (grade 3). This animal exhibits total hind-limb paralysis and urinary incontinence.

guinea pig or rat CNS tissue; human and bovine BP are relatively nonencephalitogenic for rats[11]. This seems to reflect the fact that the encephalitogenic determinant of the BP molecule which induces the disease in rats is located in a region of the molecule in which several amino acid substitutions occur, which may explain the variations in encephalitogenic activity. The encephalitogenic determinant for rats has been localized to BP residues 71-85[12,13].

Clinical signs of EAE appear approximately 9 days after injection of 25 μg guinea pig or rat BP in CFA (Fig. 2). These include loss of tail tonicity, paresis and hind limb paralysis (graded 1 to 3, respectively); by day 18 most rats recover. Mononuclear cell infiltration is particularly severe in the spinal cord (Fig. 3), but the lesions resolve with time after recovery. Cellular immunity to BP develops 7 to 9 days post-sensitization, and can be demonstrated by skin test or MIF assay[14,15].

Since Lewis rats are highly inbred, adoptive transfer experiments can be easily carried out to elucidate the cellular events involved in EAE. For example, Gonatas and Howard[16] reported that thymectomized, lethally-irradiated, bone marrow-reconstituted Lewis rats (which lack T cells but have B cell function) are resistant to EAE. This provided strong evidence that T cells are important relative to the induction of EAE.

We have found[17], as have other investigators[18,19], that 2-5 x 10^8 lymph node cells (LNC) taken 9 to 10 days after the donors were challenged with BP + CFA transfer EAE to syngeneic Lewis rat recipients. Spleen cells (SpC) do not transfer EAE in this fashion[20]. The necessity of transferring extremely large numbers of LNC has presented problems with respect to characterizing the effector cells of EAE. Moreover, the severity of disease, even in recipients of 5 x 10^8 LNC, has been quite variable, rendering quantitation difficult.

Recently, Panitch and McFarlin[21] reported that adoptive transfer of EAE in Lewis rats could be enhanced if the donor cells were cultured with the T cell mitogen concanavalin A (Con A) prior to transfer. Moreover, transfer could be achieved with Con A-cultured SpC, whereas direct transfer of SpC from donor to recipient has not been successful[20]. Subsequently, Richert et al.[22] demonstrated that donor SpC could also be activated by in vitro culture with BP. Employing the protocol depicted in Fig. 4, we confirmed this finding[23] and also observed that BP-cultured LNC readily transferred EAE (Table 1). The finding that 5 x 10^7 or fewer BP-cultured SpC will transfer severe EAE provides the opportunity to isolate and characterize the relevant effector cells.

Accordingly, we further enriched for effector cell activity by centrifugation of BP-cultured donor SpC on discontinuous Ficoll or

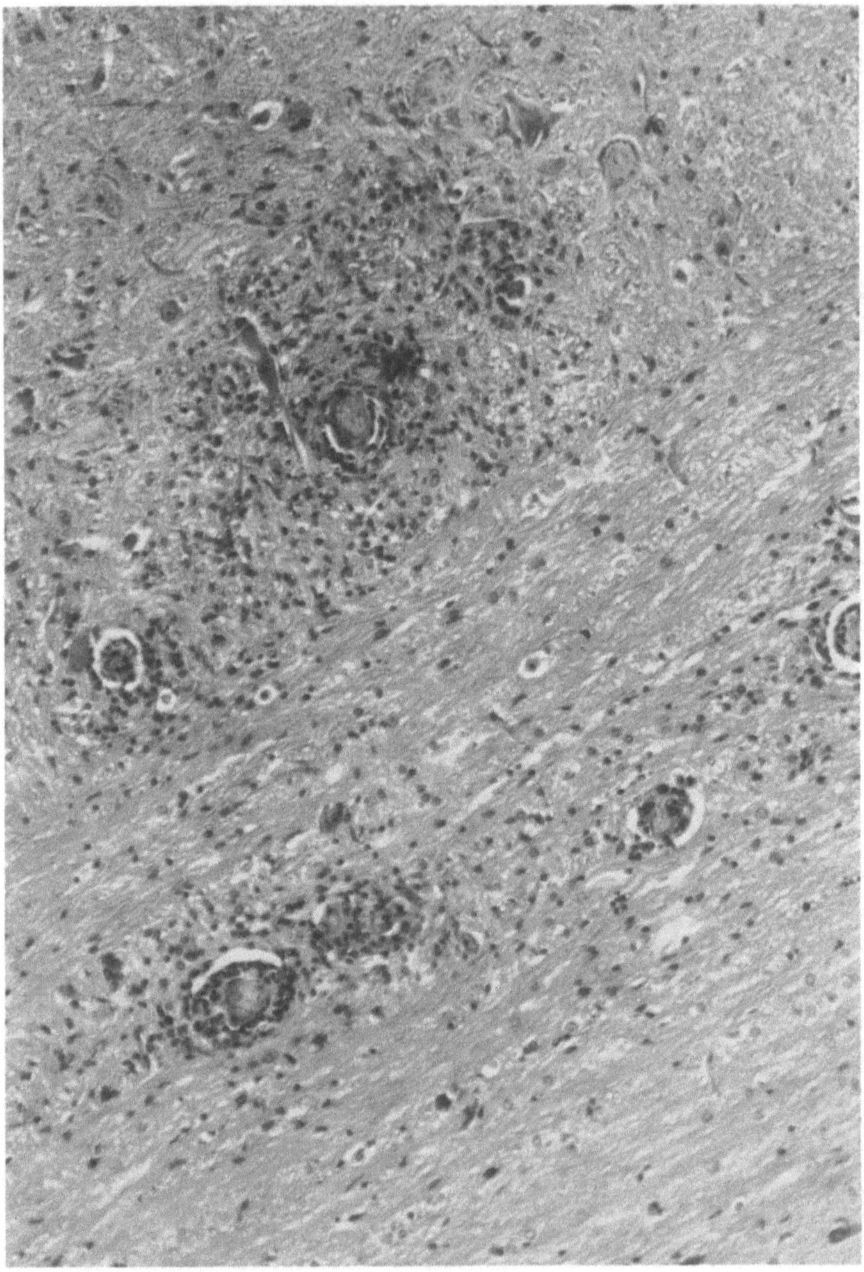

Fig. 3. Photomicrograph of the spinal cord from a rat with EAE, depicting numerous areas of mononuclear cell infiltration. Hematoxylin-eosin, X 100.

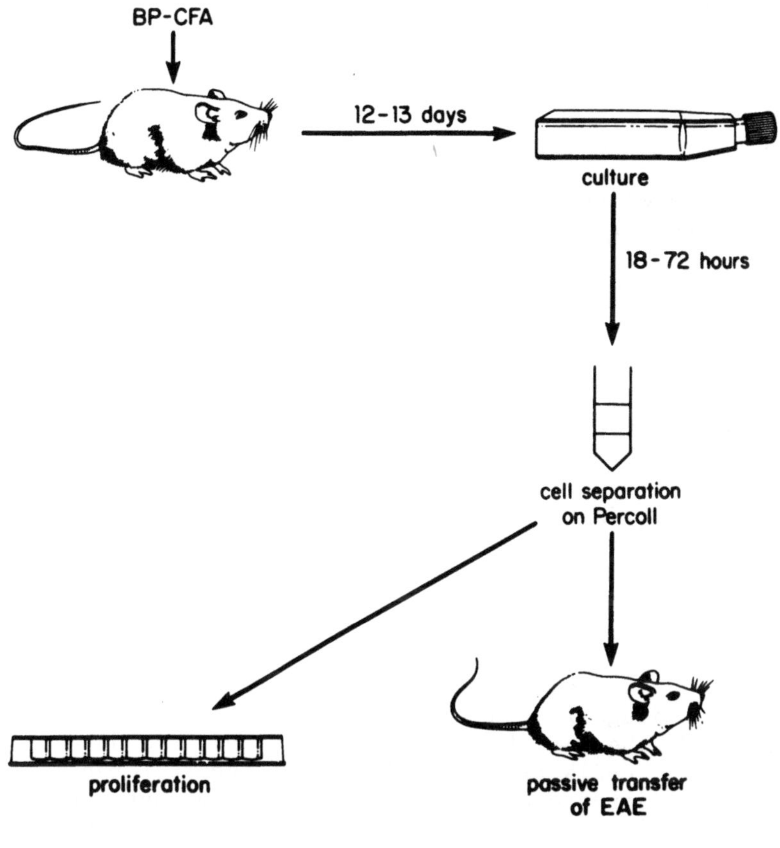

Fig. 4. Method for enhancement of EAE effector cell activity. Donor rats are challenged with BP-CFA. After 12-13 days donor SpC and LNC are cultured for 18-72 hr in RPMI 1640 + 5% fetal bovine serum + 2 μg/ml BP. The cultured cells are then washed and transferred to syngeneic recipients. Further enrichment is achieved by separation on discontinuous Percoll gradients. Proliferative assays are also carried out on aliquots of the cultured cells.

Table 1. Activation of EAE Effector Cells by *In vitro* Culture with BP

Cells Transferred[a]	Clinical EAE in Recipients Sick/Total	Days of Onset
1×10^6 SpC	3/5	4.7
5×10^7 SpC	4/4	4.5
5×10^7 LNC	6/6	3.2
Controls:		
5×10^7 SpC cultured without BP	0/8	---
7×10^7 SpC heat-killed after culture with BP	0/4	---
4×10^7 SpC from naive rats, cultured with BP	0/6	---

[a]Cells obtained 12-14 days after donor challenge, cultured 72 hr with BP, and transferred to recipients.

Percoll gradients. Initially we found that effector cells could be isolated on Ficoll gradients in the fraction that corresponded to Con-A-activated blasts run in parallel tubes (density = 1.06 g/ml). As few as 2×10^6 blast cells transferred EAE, whereas the denser small lymphocytes were inactive[23].

Subsequently, we confirmed this observation using cultured SpC isolated by centrifugation on discontinuous Percoll gradients. As shown in Table 2, the EAE effector cells floated at a density of 1.067 g/ml. These cells transferred EAE and incorporated ^{3}H-TdR in culture. Immunofluorescence analyses were carried out employing fluorescein-(FITC) conjugated rabbit anti-rat immunoglobulin (Ig) serum to detect B cells, and monoclonal anti-rat T cell serum W3/13 which labels peripheral T lymphocytes. These studies revealed that the effector cell-enriched fraction was also highly enriched for T cells (87% W3/13$^+$), and deficient in B cells (13% Ig$^+$) (Table 2). The BP-cultured EAE effector cells were also radiation-sensitive (2000 rad). On the basis of these findings, it is concluded that the effector cell that transfers EAE is a proliferating T lymphocyte.

Although EAE has been studied now for about 50 years, and is known to be mediated by T lymphocytes, the actual mechanism by which these effector cells induce EAE is still unknown. Immune tissue destruction by T cells is thought to be mediated either by lymphokine-producing or cytotoxic lymphocytes (Lyt-1 and Lyt-2,3, respectively in mice). However, evidence for T cell cytotoxicity in graft rejection is indirect, and very recent findings suggest that Lyt-1

Table 2. Characteristics of the EAE Effector Cell

Percoll density (g/ml)	Cells transferred (X10^7)	EAE in Recipients Clinical	Histologic	^{3}H-TdR Uptake (cpm)	W3/13$^+$ (%)	Ig$^+$ (%)
1.067	1-2	22/24	1.8	6285	87 $\pm$ 8	13 $\pm$ 7
1.077	1	0/12	0.1	400	46 $\pm$ 18	33 $\pm$ 4
1.090	0.5-1	0/10	0.0	80		

lymphocytes mediate rejection[24]. This would imply that lymphokines may pay a major role in T cell tissue damage, at least in the graft rejection phenomenon.

With respect to EAE, one can postulate on the basis of various pieces of information how CNS damage might come about, although clearly the puzzle is far from complete. It has been established that autoreactive T cells specific for BP transfer EAE (Table 2; ref. 23). It is also known that very few of the lymphoid cells in CNS infiltrates are of donor origin[25]. A crucial question relates to how the sensitized T cells cross the blood-brain barrier and enter the CNS. However, it may not be essential that these cells actually penetrate the blood-brain barrier. Conceivably, the initial immunologic event could occur within the blood vessels of the brain and spinal cord. Support for this notion derives from the work of Levine and Sowinski[26], who reported that the few mononuclear cells seen in the CNS of cyclophosphamide-treated rats with adoptively transferred EAE were largely confined to the walls of a few veins. If BP is catabolized and the degradation products are removed from the CNS, the BP-reactive lymphocytes could encounter the antigen in the CNS vessel lumen, localize there, and secrete chemotactic factors which attract macrophages. MIF would ensure that the macrophages were retained at the site. Sensitized T cells also produce macrophage activating factor (MAF), which stimulates production of neutral proteases. These latter enzymes could alter the permeability of the vessel walls to disrupt the blood-brain barrier and permit entry of the cells into the brain parenchyma. Cammer et al.[27] have shown that activated macrophages secrete plasminogen activator which can generate plasmin from serum plasminogen, and that plasmin can degrade BP in myelin. Moreover, Pescovitz et al.[28] reported that BP is degraded when incubated in vitro in normal serum, and suggested that trypsin-like enzymatic activity (perhaps associated with the clotting system) is responsible for proteolysis of BP. Thus, lymphokine-producing sensitized T cells may serve as the stimulus which ultimately leads to EAE, by recruiting macrophages to the CNS where the latter cause the tissue injury.

REGULATION OF EAE

Since autoreactive T cells capable of causing EAE exist, these cells must be regulated by the host to prevent the spontaneous development of autoimmune disease. Paterson and Day[29] have demonstrated myelin basic protein serum factors (MBP-SF), which appear to be products of BP catabolism. The concentration of MBP-SF varies inversely with susceptibility to EAE. For example, the serum of neonatal Lewis rats (which are resistant to EAE) contains high concentrations of MBP-SF as measured by a sensitive radioimmunoassay; as these rats mature, the concentration of MBP-SF decreases concomitant with increasing susceptibility to EAE[29]. It is postulated that MBP-SF is an endogenous neuroautotolerogen that regulates the expression of autoreactive lymphocytes having the potential for inducing EAE.

Unresponsiveness in the Guinea Pig

We have employed a somewhat different approach to study regulation of EAE. By taking advantage of the earlier findings of others that animals can be rendered unresponsive to EAE by injection of BP in nonencephalitogenic form[30], we sought to elucidate the underlying control mechanisms that regulate this autoimmune response. First, we confirmed that guinea pigs could be made unresponsive to EAE by pretreatment with multiple injections of BP in incomplete Freund's adjuvant (IFA)[31]. Emulsions of BP-IFA do not elicit EAE, nor is cellular immunity to BP induced with this material. When BP-IFA-pretreated guinea pigs were subsequently challenged with encephalitogenic BP-CFA emulsions, they did not develop EAE or cellular immunity to BP, as assayed by the MIF test (T. P. McGraw, unpublished). Unresponsiveness was specific, since guinea pigs that were similarly pretreated with HNB-BP were also unresponsive to EAE. These preparations are nonencephalitogenic for guinea pigs since HNB blocks the critical tryptophan residue in the disease-inducing site[7,8]. The finding that HNB-BP suppressed disease suggested that different molecular BP determinants were involved in the induction of EAE and unresponsiveness respectively[31]. Subsequent experiments supported this concept[32,33].

We demonstrated that unresponsiveness to EAE could be induced in guinea pigs by pretreatment with the small rat BP. This molecule lacks a sequence of 40 amino acid residues beginning just beyond the tryptophan residue and, therefore, is not encephalitogenic for guinea pigs owing to the deletion of part of the disease-inducing site[32]. Moreover, using limited pepsin digestion of BP (Fig. 1), we were able to show that BP peptide 44-89 markedly suppressed EAE in guinea pigs, further supporting the hypothesis that tolerance to EAE is a function of a suppressor determinant distinct from the encephalitogenic region of the molecule[33]. Although it is difficult to exclude the possibility that trace contamination of these preparations with intact BP contributes to unresponsiveness, we were unable to suppress EAE with the encephalitogenic peptide (corresponding to residues 114-122)[33].

In some respects this situation is analogous to hapten-carrier systems, i.e., peptide 114-122 is the hapten, and peptide 44-89 serves as the carrier. Recently we demonstrated that suppression of EAE is carrier-specific, since pretreatment with bovine serum albumin (BSA) as carrier, rendered guinea pigs unresponsive to EAE induced by encephalitogenic peptide 114-122 conjugated to BSA with 1-ethyl-3-(3-dimethyl-aminopropyl) carbodiimide. Carrier specificity is involved since BSA-pretreated animals developed EAE when challenged with BP-IFA, or with emulsions of encephalitogenic peptide (without BSA) and CFA[34]. The activities of various BP-derived preparations are summarized in Table 3.

It seemed likely that the dissociation of encephalitogenic and tolerogenic BP determinants might reflect the interaction of different lymphocyte subsets (i.e., effector and suppressor T cells) relative to this autoimmune response. Indeed, Sercarz[35] and Bullock[36] subsequently reached similar conclusions in studies with conventional antigens.

Table 3. Encephalitogenic and Tolerogenic Activity of Various BP Preparations in Hartley Guinea Pigs

Preparation	Biological Activity	
	EAE[a]	Tolerance[b]
BP	yes	yes
HNB-BP	no	yes
Small rat BP	no	yes
BP peptide 1-20	no	no
BP peptide 44-89	no	yes
BP peptide 144-122	yes	no
BP peptide 116-170	no	no
Bovine serum albumin	no	no
Bovine serum albumin; challenged with peptide 114-122-BSA	no	yes[c]

[a]Preparation injected with CFA

[b]Guinea pigs pretreated with preparation in IFA and challenged with BP-CFA (except last group; cf. footnote c).

[c]Demonstrates carrier specificity of unresponsiveness[34].

Unresponsiveness in the Lewis Rat

In order to test our hypothesis that suppressor cells may regulate EAE, we adapted the model to Lewis rats in which we could conduct cell transfer studies. We first established that, like the guinea pig, the Lewis rat can be made unresponsive to EAE by pretreatment with BP-IFA, and that unresponsiveness is immunologically specific, since irrelevant antigens do not afford protection against paralysis[17]. Pretreatment with BP peptide 44-89 also suppressed EAE in rats. This peptide also induces EAE in rats, but it has not been established whether the encephalitogenic and tolerogenic regions of this peptide are the same.

Rats rendered unresponsive with BP-IFA exhibited depressed cellular immunity to the encephalitogenic determinant when peptide 44-89 was employed as the test antigen[14]. However, cellular immunity was observed in tolerant rats when intact BP was used as the test antigen, which probably reflects responses to xenogeneic determinants present on the heterologous guinea pig BP employed routinely for induction of EAE. These findings are consistent with those of McFarlin, et al.[37], who reported that although both EAE susceptible Lewis and EAE-resistant BN rats developed delayed hypersensitivity to BP, only the Lewis strain responded to the encephalitogenic peptide.

In order to ascertain whether autoreactive lymphocytes were deleted by pretreatment with BP-IFA, we carried out adoptive transfer studies. Adoptive transfer of EAE could be routinely achieved in naive recipients with viable LNC taken from donor rats challenged 9 days earlier with BP in CFA. In contrast, when recipients were pretreated with BP-IFA prior to LNC transfer they were protected against adoptive EAE[17]. Since the LNC were washed several times before transfer, it seemed unlikely that blocking antibodies were involved.

To directly test the hypothesis that suppressor cells account for unresponsiveness to EAE, we pretreated donor rats with BP-IFA, and transferred their lymphoid cells to naive recipients which were then challenged with BP-CFA. We found that recipients of LNC or SpC from pretreated donors were protected against EAE when challenged 1 or 2 days after cell transfer, but not when challenged before cell transfer[17,38]. Thymocytes from pretreated donors did not have suppressive activity. Suppression was immunologically specific, since LNC from donors pretreated with unrelated antigens in IFA were not protected against EAE when challenged with BP-CFA. Several experiments were conducted to ascertain whether antigen carryover might account for suppression of EAE. No evidence was obtained to support such trivial explanations. For example, viable donor cells were required; killed LNC did not transfer suppression as might be expected if BP carryover was responsible[38]. Nor did normal LNC to which BP had been passively adsorbed suppress EAE in recipients[38]. LNC from pretreated donors retained suppressor activity after trypsin treatment. Such treatment might be expected to remove membrane-bound antigen, and in fact did "strip" surface immunoglobulin from B cells. Finally, surface-bound BP could not be detected by immunological or biochemical methods.

When we were reasonably confident that this was a suppressor cell phenomenon, efforts were directed toward characterizing the relevant cells. We found that suppressor activity was abrogated by treatment of the LNC with antithymocyte serum and complement. Moreover, suppressor activity was enriched in the nylon wool-non-adherent (T cell-enriched) LNC fraction (Table 4; ref. 39). These findings strongly support the conclusion that the EAE-suppressor cells in BP-IFA-treated, unresponsive rats are T lymphocytes.

Since suppressor T lymphocytes can be detected in EAE unresponsive rats, one could argue that these cells may also be present in naive animals, where they function to prevent the break-through of autoreactive effector cells which would initiate the autoimmune disease. Accordingly, EAE would be associated with a disturbance in immune homeostatis. To investigate this possibility, we have sought evidence for suppressor cells in Lewis rats challenged with BP-CFA which develop EAE and spontaneously recover from the disease. We found[40] as did Willenborg[41], that rats which have recovered from EAE are subsequently resistant to the disease regardless of whether the first episode was induced actively (with BP-CFA) or passively (with LNC from sensitized donors). However, when the first attack was elicited with SpC or LNC activated by _in vitro_ culture with BP, the recipients remained susceptible to EAE induced after recovery with BP-CFA[42]. Hinrichs et al. reported similar findings[43]. In these recipients, it was of interest that the onset of the second episode-was accelerated (approximately 7 days vs. 11 days for BP-CFA-challenged naive rats). This might suggest that the induction of EAE with BP-cultured cells bypasses recruitment of suppressor cells, which conceivably could be responsible for spontaneous recovery and subsequent resistance to the disease. Moreover, the accelerated onset in the latter animals might reflect the persistence of effector memory cells within the host.

In an effort to ascertain whether suppressor cells are associated with spontaneous recovery from EAE, we challenged donor rats with BP-CFA and transferred lymphocytes to normal recipients at various times. We found that LNC transferred prior to, or at onset of EAE in the donors, induced disease in recipients. Donor LNC transferred after recovery (i.e., > 21 days post-challenge) were ineffective with respect to transfer of EAE. In contrast, these cells exhibited suppressor activity, since the recipients were protected against EAE when challenged with BP-CFA[40]. Representative results are shown in Table 5, which also reveals that these "post-

Table 4. Characterization of EAE Suppressor Cells in BP-IFA-Treated Rats

Suppressor Cell Treatment	Cells Transferred ($X\ 10^8$)	Clinical EAE in Recipients
Antithymocyte serum[a] and Complement	5.9	5/5[b]
Normal Rabbit serum and Complement (Control)	4.8	0/3
Nylon Wool Nonadherent	2.4-3.4	0/6

[a]Prepared in rabbits
[b]Number sick/total

recovery" suppressor cells differ from those found in tolerant Lewis rats (See Table 4), the former being adherent to nylon wool.

Although the postrecovery suppressor cells are adherent to nylon wool, they do not appear to be macrophages[40]. It is not clear at present whether they are adherent T cells or B cells which produce anti-BP blocking antibody. However, attempts to suppress EAE with serum from recovered rats have not been successful, suggesting that circulating antibodies are not involved in suppression (J. Killen, unpublished).

The post recovery suppressor cells seem to function at the afferent phase of the immune response, since they will not inhibit transfer of EAE by *in vitro*-activated SpC from sensitized donors (Table 6).

Thus we conclude that the appearance of suppressor cells correlates with spontaneous recovery from EAE, although their precise role in recovery must still be established.

It is clear from our previously published reports that unresponsive, suppressor cell-bearing Lewis rats (both BP-IFA-treated tolerant rats[44] and nonsusceptible recovered rats[23]) also harbor effector cells which can be activated by *in vitro* culture with BP. Representative results are shown in Table 7. However, effector cell activity is not demonstrable if lymphocytes from tolerant donors are transferred directly without *in vitro* culture.

The demonstration of effector cells, despite the fact that the hosts are not susceptible to EAE, argues in favor of active immuno-

Table 5. Characterization of Postrecovery Suppressor Cells Transferred 30 Days After Donor Challenge with BP-CFA

Cells Transferred	Clinical EAE in Recipients[a] Severity[b]
5 x 10^8 unfractionated LNC	0.5
1 x 10^8 nylon adherent LNC	0.8
1 x 10^8 nylon nonadherent LNC	2.7

[a]Recipient challenged 8 days post-transfer
[b]Mean group severity (max. = 3.0)

Table 6. Inability of Postrecovery Suppressor Cells to Inhibit Transfer of EAE by BP-activated SpC from Donors Challenged with BP-CFA 12 Days Previously

Recipient Treatment	EAE in Recipients (Severity)
5 x 10^8 postrecovery suppressor cells[a] + BP-CFA on day 8	0.5
5 x 10^8 postrecovery suppressor cells[a] + 5 x 10^7 BP-activated SpC on day 8	3.0

[a]LNC transferred 30 days post challenge

regulation of this autoimmune response rather than depletion of immunocompetent effector cells. We postulated that host suppressor cells mediate this regulation of autoimmunity, although other mechanisms (e.g., receptor blockade by tolerogenic self-antigen; ref. 29,45) cannot be excluded.

CONCLUSIONS

Previous studies of the immunochemistry of myelin basic protein and the immunobiology of EAE have provided considerable insight into mechanisms governing immune effector function and regulation. With respect to the CNS, EAE continues to serve as a model for human

Table 7. Demonstration of EAE Effector SpC in Unresponsive Lewis Rats

SpC Donor[a]	Clinical EAE in Recipients Sick/Total
Recovered from active EAE	4/4
BP-IFA-treated, BP-CFA challenged (tolerant)	6/7

[a]Donor SpC activated culturing with BP for 72 hours; 10^7 cells transferred/recipient.

demyelinating diseases[1]. It is intriguing that recent evidence points toward a possible defect in suppressor cells in MS patients[46]. By analogy with EAE, one might speculate that the clinical manifestations of MS relate to an immunologic attack against the CNS which is provoked when normal regulatory mechanisms fail. The immunologic attack could conceivably involve autoreactive lymphocytes specific for CNS antigens. Alternatively, the immunologic CNS damage could be nonspecific, caused by mediators released from cells which are reactive against foreign antigens (e.g., viruses), and which accumulate in the CNS merely because this is the site where the antigen is localized. Evidence for this sort of "bystander" effect has accrued from several studies which suggest that cellular immunity to viruses[47], or to tuberculoprotein antigens[48] within the CNS can have immunopathologic consequences. The effect on the host would be the same regardless of whether the lymphocytes that entered the CNS were autoreactive or directed against exogenous antigens. In either case, products released from the CNS lymphocytes or macrophages subsequently brought into the lesions would cause demyelination. Nevertheless, it would be beneficial to ascertain the immunological specificity of the CNS lymphocytes in order to better understand the etiology and more effectively design appropriate therapeutic regimens.

ACKNOWLEDGEMENTS

The studies carried out in the authors' laboratory were supported by NIH research grant NS-06985-14, and National Multiple Sclerosis Society grant RG 1073-B-5. James Holda and Joyce Killen were supported in part by NIH training grant 1T32-AI 07118. We thank Mr. Keith Varhol and Ms. Kim Martin for expert technical assistance, and Ms. Chuty Erves for excellent secretarial help.

REFERENCES

1. P. Y. Paterson, Experimental autoimmune encephalomyelitis: induction, pathogenesis, and suppression, in: "Textbook of Immunopathology," P. A. Miescher and H. J. Muller-Eberhard, eds., Grune and Stratton, New York (1976), p. 179.
2. N. R. Rose, Autoimmune diseases, Scientific American, 244:80 (1981).
3. R. H. Swanborg, J. E. Swierkosz, and R. G. Saieg, Studies on the species variability of experimental allergic encephalomyelitis in guinea pigs and rats, J. Immunol. 112:594 (1974).
4. E. H. Eylar, Amino acid sequence of the basic protein of the myelin membrane, Proc. Nat. Acad. Sci. 67:1425 (1970).
5. P. R. Carnegie, Amino acid sequence of the encephalitogenic basic protein from human myelin, Biochem. J. 123:57 (1971).
6. R. E. Martenson, G. E. Deibler, M. W. Kies, S. S. McKneally, R. Shapira, and R. F. Kibler, Differences between the two myelin basic proteins of the rat central nervous system: a deletion in the smaller protein. Biochim. Biophys. Acta. 263:193 (1972).
7. R. H. Swanborg, The effect of selective modification of tryptophan, lysine and arginine residues of basic brain protein on encephalitogenic activity, J. Immunol. 105:865 (1970).
8. L. P. Chao and E. R. Einstein, Localization of the active site through chemical modification of the encephalitogenic protein, J. Biol. Chem. 245:6397 (1970).
9. C-M. Shaw, E. C. Alvord, Jr., J. Kaku, and M. W. Kies, Correlation of experimental allergic encephalomyelitis with delayed-type skin sensitivity to specific homologous encephalitogen, Ann. N. Y. Acad. Sci. 122:318 (1965).
10. R. P. Lisak and B. Zweiman, In vitro and in vivo immune responses to homologous myelin basic protein in experimental allergic encephalomyelitis, Cell. Immunol. 11:212 (1974).
11. R. E. Martenson, G. E. Deibler, M. W. Kies, S. Levine, and E. C. Alvord, Jr., Myelin basic proteins of mammalian and submammalian vertebrates: encephalitogenic activities in guinea pigs and rats, J. Immunol. 109:262 (1972).
12. C-H. J. Chou, R. B. Fritz, F. C-H. Chou, and R. F. Kibler, The immune response of Lewis rats to peptide 66-88 of guinea pig myelin basic protein. I. T cell determinants, J. Immunol. 123:1540 (1979).
13. G. A. Hashim, Myelin basic protein: structure, function and antigenic determinants, Immunol. Rev. 39:60 (1978).
14. T. P. McGraw and R. H. Swanborg, Cell-mediated immunity to myelin basic protein in Lewis rats made unresponsive to experimental allergic encephalomyelitis, Eur. J. Immunol. 8:905 (1978).
15. F. J. Waxman, R. B. Fritz, and D. J. Hinrichs, The presence of specific antigen-reactive cells during the induction,

recovery, and resistance phases of experimental allergic encephalomyelitis, Cell. Immunol. 49:34 (1980).

16. N. K. Gonatas and J. C. Howard, Inhibition of experimental allergic encephalomyelitis in rats severely depleted of T cells, Science. 186:839 (1974).
17. J. E. Swierkosz and R. H. Swanborg, Suppressor cell control of unresponsiveness to experimental allergic encephalomyelitis, J. Immunol. 115:631 (1975).
18. P. Y. Paterson, Passive transfer of allergic encephalomyelitis in rats by means of lymph node cells, J. Exp. Med. 111:119 (1960).
19. S. Levine, E. M. Hoenig, and M. W. Kies, Allergic encephalomyelitis: immunologically specific inhibition of cellular passive transfer by encephalitogenic basic protein, Clin. Exp. Immunol. 6:503 (1970)
20. P. Y. Paterson, Experimental allergic encephalomyelitis and autoimmune disease, Advanc. Immunol. 5:131 (1966).
21. H. S. Panitch and D. E. McFarlin, Experimental allergic encephalomyelitis: enhancement of cell-mediated transfer by concanavalin A, J. Immunol. 119:1134 (1977).
22. J. R. Richert, B. F. Driscoll, M. W. Kies, and E. C. Alvord, Jr., Adoptive transfer of experimental allergic encephalomyelitis: incubation of rat spleen cells with specific antigens, J. Immunol. 122:494 (1979).
23. J. H. Holda, A. M. Welch, and R. H. Swanborg, Autoimmune effector cells. I. Transfer of experimental allergic encephalomyelitis with lymphoid cells cultured with antigen, Eur. J. Immunol. 10:657 (1980).
24. B. E. Loveland, P. M. Hogarth, R. Ceredig, and I. F. C. McKenzie, Cells mediating graft rejection in the mouse. I. Lyt-1 cells mediate skin graft rejection, J. Exp. Med. 153:1044 (1981).
25. O. Werdelin and R. T. McCluskey, The nature and specificity of mononuclear cells in experimental autoimmune inflammations and the mechanisms leading to their accumulation, J. Exp. Med. 133:1242 (1971).
26. S. Levine and R. Sowinski, Allergic encephalomyelitis: new form featuring polymorphonuclear leukocytes, Science. 171: 498 (1971).
27. W. Cammer, B. R. Bloom, W. T. Norton, and S. Gordon, Degradation of basic protein in myelin by neutral proteases secreted by stimulated macrophages: a possible mechanism of inflammatory demyelination, Proc. Nat. Acad. Sci. 75: 1554 (1978).
28. M. D. Pescovitz, P. Y. Paterson, J. Kelly, and L. Lorand, Serum degradation of myelin basic protein with loss of encephalitogenic activity: evidence for an enzymatic process, Cell. Immunol. 39:355 (1978).
29. P. Y. Paterson, E. D. Day, and C. C. Whitacre, Neuroimmunologic diseases: effector cell responses and immunoregulatory

mechanisms, Immunol. Rev. 55:89 (1981).
30. E. C. Alvord, Jr., C-M. Shaw, S. Hruby, and M. W. Kies, Encephalitogen-induced inhibition of experimental allergic encephalomyelitis: prevention, suppression and therapy, Ann. N. Y. Acad. Sci. 122:333 (1965).
31. R. H. Swanborg, Antigen-induced inhibition of experimental allergic encephalomyelitis. I. Inhibition in guinea pigs injected with nonencephalitogenic modified myelin basic protein, J. Immunol. 109:540 (1972).
32. R. H. Swanborg, Antigen-induced inhibition of experimental allergic encephalomyelitis. II. Studies in guinea pigs with the small rat myelin basic protein, J. Immunol. 111: 1067 (1973).
33. R. H. Swanborg, Antigen-induced inhibition of experimental allergic encephalomyelitis. III. Localization of an inhibitory site distinct from the major encephalitogenic determinant of myelin basic protein, J. Immunol. 144:191 (1975).
34. H. C. Rauch, I. N. Montgomery, and R. H. Swanborg, Inhibition of experimental allergic encephalomyelitis by carrier administered prior to challenge with encephalitogenic peptide-carrier conjugate, Eur. J. Immunol. 11:335 (1981).
35. D. Turkin and E. E. Sercarz, Key antigenic determinants in regulation of the immune response, Proc. Nat. Acad. Sci. 74:3984 (1977).
36. W. W. Bullock, ABA-T determinant regulation of delayed hypersensitivity, Immunol. Rev. 39:3 (1978).
37. D. E. McFarlin, S. C-L. Hsu, S. B. Slemenda, F. C-H. Chou, and R. F. Kibler, The immune response against myelin basic protein in two strains of rat with different capacity to develop experimental allergic encephalomyelitis, J. Immunol. 141:72 (1975).
38. J. E. Swierkosz and R. H. Swanborg, Immunoregulation of experimental allergic encephalomyelitis: conditions for induction of suppressor cells and analyses of mechanism, J. Immunol. 119:1501 (1977).
39. A. M. Welch and R. H. Swanborg, Characterization of suppressor cells involved in regulation of experimental allergic encephalomyelitis, Eur. J. Immunol. 6:910 (1976).
40. A. M. Welch, J. H. Holda, and R. H. Swanborg, Regulation of experimental allergic encephalomyelitis. II. Appearance of suppressor cells during the remission phase of the disease, J. Immunol. 125:186 (1980).
41. D. O. Willenborg, Experimental allergic encephalomyelitis in the Lewis rat: studies on the mechanism of recovery from disease and acquired resistance to reinduction, J. Immunol. 123:1145 (1979).
42. J. H. Holda and R. H. Swanborg, Susceptibility of Lewis rats to experimental autoimmune encephalomyelitis after recovery from passively induced disease, Immunol. Commun. 9:333 (1980).

43. D. J. Hinrichs, C. M. Roberts, and F. J. Waxman, Regulation of paralytic experimental allergic encephalomyelitis in rats: susceptibility to active and passive disease reinduction, J. Immunol. 126:1857 (1981).
44. J. H. Holda and R. H. Swanborg, Regulation of experimental allergic encephalomyelitis. III. Demonstration of effector cells in tolerant rats, Eur. J. Immunol. 11:338 (1981).
45. I. R. Cohen and H. Wekerle, Regulation of autosensitization: the immune activation and specific inhibition of self-recognizing thymus-derived lymphocytes, J. Exp. Med. 137:224 (1981).
46. J. P. Antel, B. G. W. Arnason, and M. E. Medof, Suppressor cell function in multiple sclerosis: correlation with clinical disease activity, Ann. Neurol. 5:338 (1979).
47. M. C. Dal Canto and S. G. Rabinowitz, Central nervous system demyelination in Venezuelan equine encephalomyelitis infection: an experimental model of virus-induced myelin injury, J. Neurol. Sci. 49:397 (1981).
48. H. M. Wisniewski and B. R. Bloom, Primary demyelination as a nonspecific consequence of a cell-mediated immune reaction, J. Exp. Med. 141:346 (1975).

2

THE USE OF HYBRIDOMA ANTIBODIES TO PROBE THE ANTIGENIC DETERMINANTS OF THYROGLOBULIN

N. R. Rose, M. Accavitti, E. F. Pydyn, M. A. Leon and R. K. Brown

Wayne State University School of Medicine, Department of Immunology and Microbiology and the Department of Biochemistry, 540 East Canfield Avenue, Detroit, Michigan 48201

INTRODUCTION

Experimental autoimmune thyroiditis has provided a fruitful model for the study of human autoimmune disease and for deepening our understanding of the mechanisms of self-non-self discrimination[1]. The disease is produced by injection of mice (or other experimental animals) with crude mouse thyroid extract or purified mouse thyroglobulin (MTg) together with a suitable adjuvant. Injection of complete Freund's adjuvant (CFA) with MTg elicits production of autoantibodies to thyroglobulin and infiltration of the thyroid gland by lymphocytes, macrophages and other mononuclear cells; i.e., chronic thyroiditis is produced. Injection of large doses of MTg without adjuvant renders mice unresponsive to MTg, due to production of MTg-specific suppressor T cells[9]. Immunization with Tg under diverse conditions, therefore, may lead to autoantibodies alone, to both autoantibodies and thyroid lesions, or to suppression.

Genetic regulation of EAT

When many inbred strains of mice were tested for their responsiveness to thyroglobulin, significant differences were found[2]. Although all mice eventually produced autoantibodies after injection of MTg and CFA, some strains had detectable antibody within a week while other strains required several weeks to produce measurable levels of autoantibody[3]. The prompt responders usually developed more severe inflammation of thyroid with massive mononuclear infiltration and disruption of the normal architecture of the thyroid gland, whereas the late responders had little or no infiltration of their thyroids. More recently we have used bacterial lipopolysaccharide (LPS), a weaker adjuvant, and been able to make an even clearer distinction between good responders and poor responders with respect to antibody levels and thyroid pathology[4].

In our initial studies, over 30 different strains of mice were classified as good responders or poor responders based on the criteria discussed above[2]. The classifications hold for both adjuvants. showing that the autoimmune response to MTg, and not susceptibility to the adjuvant, was under genetic regulation[3]. Furthermore, the distinction between responder and non-responder strains was the same regardless of the dose (from 20 to 200 μg per injection) or the strain of origin of MTg[5]. All mouse strains carrying the H-2 haplotype q, s or k were good responders while all mice with H-2 haplotypes b, d or f were poor responders[3]. Examination of congenic strains that differ only at H-2 confirmed the conclusion that the most important gene(s) controlling this autoimmune response was located within the H-2 complex[2].

By the use of intra-H-2 recombinants we further localized the major immunoregulatory (Ir) gene controlling response to MTg (referred to as Ig-Tg) to the K end of the H-2 complex or the I-A subregion[6]. More recent investigations with newly developed recombinants have shown definitively that Ir-Tg is in the I-A subregion[7].

In the course of studies on intra-H-2 recombinants, we observed that strains with good responder alleles at I-A but various other alleles to the left of the I-B subregion promptly developed high titers of thyroglobulin autoantibody but had only moderate lesions[6]. The best explanation was that there is a separate locus modifying cytotoxic effects located nearer the D end of the H-2 complex. This prediction has recently been verified by the demonstration of effector control at the D region[8]. Moreover, alleles in the K locus also modify effector reactions[7].

A picture emerges, then, of a double level of genetic regulation of the autoimmune response to MTg. The initial recognition of the molecule is probably determined by the balance of helper versus suppressor T cells stimulated by various epitopes. This step of the response is regulated by Ir-Tg, an immunoregulatory gene at I-A. In animals that are good responders, the severity of disease is modified by other genes at the D or K regions.

Genes outside of H-2 also have impact on the autoimmune response to thyroglobulin. Examination of a large panel of identical H-2 haplotypes on different backgrounds has shown significant differences in antibody titer and thyroid pathology[9]. Some of these non-H-2 genes may be linked to minor histocompatibility genes or to genes regulating immunoglobulin synthesis.

Thyroglobulin, a multideterminant autoantigen

Thyroglobulin is a large glycoprotein with a molecular weight of about 660,000 daltons which is composed of two similar subunits. It has been estimated that when human thyroglobulin is injected into a foreign species like a rabbit, it expresses 30 or 40 different epitopes. In contrast, human autoantibodies recognize only 4 to 6 determinants on this molecule, suggesting that only about one tenth of the potential antigenic determinants are autoantigenic[10,11].

Previous investigators have attempted to degrade the thyroglobulin molecule into small molecular weight peptides bearing individual antigenic determinants. Metzger et al.[12] isolated tryptic fragments of Tg with an estimated molecular weight of 700 which partially inhibited a heterologous Tg immune system. Larger fragments with a molecular weight of about 8000 were also prepared which completely inhibited. The fragments were not further characterized.

Stylos and Rose[13] studied digests of human thyroglobulin made with a variety of proteinases. Tryptic fragments which precipitated with rabbit anti-Tg (i.e., heteroantibodies) and not with human anti-Tg (i.e., autoantibodies) were obtained. They gave weak hemagglutination inhibition in both rabbit and human systems. Chymotrypsin gave no immunoreactive fragments. Two papain fragments were partially purified. One reacted with rabbit but not human antibody. The second, smaller fragment did not precipitate with either antiserum but was a weak inhibitor of hemagglutination in both systems. The pepsin fragments obtained precipitated with the rabbit or human anti-Tg and strongly inhibited hemagglutination in both systems. Two pepsin fragments were partially purified by gel filtration[14]. Both reacted with hetero- and autoantibodies. One fragment approached the activity (on a weight basis) of Tg in hemagglutination inhibition. The second smaller fragment was as effective (on a weight basis) as Tg in inhibiting the heteroimmune system but only 1/16 as active in the autoimmune system. Similar observations were recently reported by Nye et al.[11].

In another study, sialic acid, the terminal sugar in one of the two carbohydrate units, was removed, resulting in enhancement of immunoreactivity of Tg. Desialation may, therefore, unmask antigenic determinants[15]. The glycopeptides obtained by pronase digestion inhibited the reaction between human Tg and autoantibody[15]. Although the glycopeptides were not characterized as to amino acid composition, they were prepared by the methods which yield chiefly the polysaccharides terminated with a single asparagine. Other studies have examined the role of thyroxine[16], iodine, or sugar residues in antigenic determinants[17]. In general, however, the quoted studies have used heteroimmunization and their applicability to autoimmune responses is not known.

Other chemical properties of Tg that are well documented are: amino acid composition,[18,19] the structure of the two carbohydrate units A and B,[20,21,22,23] and properties of carbohydrate-stripped Tg[24,25].

We have now turned our attention to those sequential and conformational determinants of the MTg molecule that control helper and suppressor functions in murine autoimmune thyroiditis. Studies of antigenic determinants of proteins have, in the past, concerned primarily those structures that are involved in reaction with antibodies. Two proteins, myoglobin and lysozyme, have been extensively studied from this point of view[26,27]. Only recently, based on Atassi's studies, are antigenic determinants being examined for their ability to induce helper or suppressor cells[28,29]. Our work also has analogies to studies in acute allergic encephalomyelitis which have linked induction of disease to specific amino acid sequences of myelin basic protein[30,31] and are now dissecting the role of suppressor T cells[32]. As in our previous studies of thyroid alloantigens[37], we have utilized monoclonal antibodies, from the mouse's own repertoire, to define autoantigenic determinants and to study the role of these autoantigenic determinants in autoimmune thyroiditis.

Considerations of the size and complexity of the MTg molecule had initially suggested that our efforts should concentrate on isolation of fragments that are deficient in one or more antigenic determinants as defined by our library of monoclonal antibodies to MTg. However, in preliminary experiments, we have been greatly impressed with both the speed and resolving power of preparative high pressure liquid chromatography (HPLC). We are now more optimistic that the antigenic determinants of MTg responsible for its autoantigenic properties can be isolated and identified.

Production of monoclonal antibodies to murine Tg

We have cloned 20 hybridoma lines producing anti-MTg of which 18 have been characterized extensively. The lines were obtained by fusing BALB/c plasmacytoma lines, P3X20, NS-1 or P3X63 (6.53) with spleen cells of good responder (C3H/Anf x BALB/c)F_1 mice immunized to MTg. The hybridoma products were screened for their ability to agglutinate erythrocytes coated with MTg by the chromium chloride technique. Reactive cell lines included hybridomas producing IgM, IgA, IgG_1, IgG_{2a} and IgG_{2b}. The antibodies were purified by elution from immunoadsorbent columns of MTg coupled to Sepharose 4B using 2M thiocyanate as eluant.

Reactions of monoclonal antibodies to murine Tg

If a hybridoma IgG antibody reacts with a unique antigenic determinant of a monomeric protein, then the only complexes which can form have the composition ag-ab or ag_2-ab. No lattice formation is possible and under normal circumstances such complexes should not precipitate. However, if the protein exists in solution as a dimer, then depending on the relative concentrations of antigen and monoclonal antibody in solution, linear complexes of antigen and antibody of various sizes can form, viz., -ag.ag-ab-ag.ag-ab-, etc. Although no three dimensional lattice can form, depending on the physico-chemical properties of the complex, precipitation may or may not take place. Finally, if the protein exists in solution as a tetramer and if the antigenic determinant on each subunit of the tetramer is available to antibody, then reaction with monoclonal antibody should result in a lattice with precipitation.

```
-ab-ag.ag-ab-
     .  .
-ab-ag.ag-ab-
```

In the case of IgM hybridoma antibodies to monomeric proteins, where theoretically the complex $IgM\text{-}ag_{10}$ can form, precipitation would depend on the extent of steric hindrance and the particular physico-chemical characteristics of that complex. However, dimers should be capable of lattice formation with IgM hybridoma antibody and therefore should precipitate. By analogy with the structure of human Tg, MTg is at least a dimer, so all IgM hybridomas should precipitate MTg unless unusual steric factors interfere.

Our experience with cloned hybridoma antibodies to MTg has demonstrated that no IgG antibodies precipitate, 1 of 2 IgA antibodies precipitates and 2 of 2 IgM antibodies precipitate. Tests were carried out by the Ouchterlony procedure. Electrophoresis on 5% gels under non-denaturing conditions demonstrated that the precipitating IgA was isolated mainly as dimer while the non-precipitating IgA was present almost exclusively as monomer. Several of the non-precipitating IgG antibodies showed precipitation with MTg when $CdSO_4$, an enhancer of precipitation[33] was added to the gel. We suspect these to be instances where the linear complexes discussed above are of sufficient size to precipitate when $CdSO_4$ is present. When individual non-precipitating thyroglobulin antibodies were mixed with each other, two at a time, and diffused against Tg, in the presence of $CdSO_4$, many instances of precipitation were observed. We term such precipitations, resulting from mixing two monoclonal antibodies, as complementation, and predict that such complementation occurs when the two antibodies are reacting with two distinct antigenic determinants of a dimer (D_1 and D_2) that do not interfere sterically as shown below:

$$\begin{matrix}-ab_1-\\-ab_1-\end{matrix}\begin{bmatrix}D_1.D_2\\D_1.D_2\end{bmatrix}\begin{matrix}-ab_2-\\-ab_2-\end{matrix}\begin{bmatrix}D_2.D_1\\D_2.D_1\end{bmatrix}\begin{matrix}-ab_1-\\-ab_1-\end{matrix}\begin{bmatrix}D_1.D_2\\D_1.D_2\end{bmatrix}\begin{matrix}-ab_2-\\-ab_2-\end{matrix}$$

In practice, some of the IgG monoclonals, when reacted with MTg, produce a localized decrease in the opacity of the agar. This appearance is referred to as a "negative line" and recalls a similar phenomenon described many years ago by Goudie et al.[34] and by our group[35]. Negative lines remain in agar even after thorough washing of the plates in saline solution, and can be stained by azocarmine. We assume that they are due to the localization of linear antigen-antibody complexes described above.

The complementation test provides a means for estimating the minimal number of antigenic determinants on a molecule as well as nearest neighbors. The larger the population of hybridoma antibodies examined, the closer this minimal number will approach the actual number of antigenic determinants recognized by the animal's immune system.

The results of the complementation studies (Table 1) indicate that 11 of the hybridoma IgG antibodies have unique complementation

Table 1. Complementation Patterns for Non-Precipitating Hybridoma Antibodies

	BS24	D16	D3	RKB	AK22	D23	JS6	NR	MAL	D10	CF14	PA18	MMA3	AN24	EP11
BS24	-	0	0	X	X	X	X	X	X	X	X	X	X	X	X
D16	0	-	0	X	X	X	X	X	X	X	X	X	X	X	X
D3	0	0	-	X	X	X	X	X	X	X	X	X	X	X	X
RKB	X	X	X	-	X	X	X	X	X	X	X	0	0	X	X
AK22	X	X	X	X	-	0	0	X	X	X	X	X	0	0	X
D23	X	X	X	X	0	-	0	X	X	X	X	X	0	X	X
JS6	X	X	X	X	0	0	-	X	X	X	X	X	0	X	X
NR	X	X	X	X	X	X	X	-	0	0	0	X	X	X	X
MAL	X	X	X	X	X	X	X	0	-	0	0	X	X	X	0
D10	X	X	X	X	X	X	X	0	0	-	0	0	X	0	X
CF14	X	X	X	X	X	X	X	0	0	0	-	0	X	0	0
PA18	X	X	X	0	X	X	X	X	X	0	0	-	0	0	0
MMA3	X	X	X	0	0	0	0	X	X	X	X	0	-	0	0
AN24	X	X	X	X	0	X	X	X	X	0	0	0	0	-	0
EP11	X	X	X	X	X	X	X	X	0	X	0	0	0	0	-

X = precipitation (complementation)
0 = no complementation

patterns, with identical patterns in two instances (JS6 = D23; D16 = D3 = BS24). AK22 is very closely related to D3 and JS6. CF14, D10, MAL and NR form another closely related family. On this basis the mouse recognizes a minimum of 8 different autoantigenic determinants on the Tg molecule. In Table 1 those antibodies with complementation patterns that are closely related are grouped together within horizontal lines. Since the antibodies in any group do not complement each other we believe that they either recognize portions of the same determinant or neighboring antigenic determinants.

While our studies on complementation by non-precipitating hybridomas were in progress, another group reported that combinations of 2 non-precipitating monoclonal antibodies may form precipitates in the presence of PEG 3000[36]. They utilized turbidimetric methods to study complex formation between IgG and monoclonal anti-IgG and by zonal centrifugation measured the sedimentation coefficients of the complexes generated. They noted that one combination of anti-Cγ3 antibodies gave only a trace of turbidity with PEG 3000. We suspect that this result is due to steric interference i.e. two antigenic determinants that are near neighbors may not form a precipitate in the presence of the monoclonal antibodies to these determinants. Two distant determinants generally should do so.

Additional information on differences in specificity of the hybridoma antibodies was obtained by examining their ability to cross-react with rat, rabbit, bovine, or porcine Tg. The results, shown in Table 2, were obtained by hemagglutination of erythrocytes coated with the respective Tg. It is clear that cross-reactivity represents another parameter by which monoclonal antibodies can be differentiated. Of particular interest is the pattern of cross-reactivity of hybridoma RKB with rat, rabbit, bovine, and porcine Tg. Such a highly conserved antigenic determinant may play a significant role in the biological activity of Tg. The presence of this determinant in bovine Tg, a commercially available protein, made its isolation an immediate goal (vide infra).

The data of Table 1 show that monoclonal antibodies D16, BS24 and D3 complemented the same group of hybridoma antibodies and did not complement each other. On the basis of these studies, we would argue that they either recognized the same determinant, portions of the same determinant, or different determinants that were spatially so close together that steric effects prevented both antibodies from reacting and forming a lattice. That the first interpretation is incorrect for BS24 and D3 is shown by the cross-reactivity data in Table 2 which clearly distinguish between the specificities of the two antibodies. One cannot draw the same conclusion with respect to D16 since it may be of lower affinity than either BS24 and D3 and therefore shows diminished cross reactivity. D23 and AK22, which show nearly identical patterns in cross-reactivity studies, and do not complement each other, may recognize the same determinant,

Table 2. Reactivity of Monoclonal Antibodies with Erythrocytes Coated with Tg from Various Species

	Ab Class	Parental Line	MOUSE	RAT	RABBIT	PORCINE	BOVINE
MMA3	G2a,k	P3X20	22+	0	0	0	0
CF14	G1,k	P3X20	22	10	1	0	0
BC7	IgMk	P3X20	22	0	0	0	0
AN24	G2b,k	NS-1	22	12*	0	0	0
D3	G1,k	NS-1	23	12*	0	0	0
D23	G1,k	NS-1	23	23	0	9	7
D14	IgA,k	NS-1	24	12*	0	6	0
RKB	G2a,k	P3X63(6.53)	23	23*	23*	23*	23*
NR	G1,k	P3X63(6.53)	23	12*	0	0	0
MAL	G1,k	P3X63(6.53)	22	12*	0	0	0
PA18	IgA,k	P3X63(6.53)	25	23	0	12*	12*
D10	G1,k	P3X63(6.53)	22	12*	2	4	1
D16	G1,k	P3X63(6.53)	20	3	0	0	0
JS6	G1,k	P3X63(6.53)	23	8	0	0	1
AK22	G1,k	P3X63(6.53)	28	23	2	6	7
AR18	IgM,k	P3X63(6.53)	22	1	1	2	0
EP11	G1,k	P3x63(6.53)	23	12*	0	0	1
BS24	G1,k	P3X63(6.53)	22	7	9	10	10

+ log 2 titers

* Not tested at higher dilutions

JS6, nearly identical in complementation to AK22, differs markedly in its pattern of cross-reactivity.

The third parameter that we have used to distinguish the various clones is isoelectric focusing (IEF). IEF of the hybridomas using commercial ampholytes gave poor results in our system. We, therefore, synthesized our own ampholytes from pentaethylenehexamine and prepared a mixture that gave good patterns with both the IgG and IgA monoclonal antibodies. The results showed distinct focusing patterns for each IgG and IgA antibody including AK22 and D23 (both $IgG_{1,k}$). Considered alone, these IEF data provide no information as to whether these physico-chemically unique clones also have unique antigenic specificities. Taking all three parameters together, IEF, complementation and cross-reactivity, we conclude that each of the hybridoma antibodies is unique, either with respect to some detail of structure or antigenic specificity. How the uniqueness defined in these tests would be reflected by anti-idiotype analysis is as yet unknown. However, we predict that the resolving power of the battery of tests used above is greater than that of conventional heterogeneous anti-idiotypic sera.

We wish to emphasize that the isolation of many unique monoclonal antibodies does not imply that the mouse normally responds to MTg with similar amounts of all of these antibodies. Some of them may represent rare idiotypes without significant biological roles. Similarly, the number of antigenic determinants involved in regulation of autoimmune thyroiditis may be far smaller than the repertoire defined by our hybridoma antibodies.

Isolation of biologically active fragments

The monoclonal antibodies to MTg also define evolutionarily conserved sequences in heterologous Tgs. As shown in Table 2 some heterologous Tgs share a limited number of determinants, similar to those on MTg, and therefore may be used as models for antigenically deficient fragments of MTg. The greater availability of some heterologous Tgs as compared to MTg makes them particularly useful. Since several hybridoma antibodies to MTg cross-react with bovine Tg, we have begun to isolate the cross-reactive determinant(s) to determine whether evolutionary conservation is associated with an important biological role. Small fragments of bovine Tg were prepared by digestion with TPCK treated trypsin in 0.1 M NH_4HCO_3 for 4 hr at 37°. The digest was sieved on a column of Sephadex G-50 superfine and the fractions lyophilized, reconstituted in phosphate buffered saline and tested by hemagglutination inhibition. Three hybridoma antibodies, RKB, BS24 and PA18, which cross-reacted with bovine Tg were studied. All were inhibited by fragments as small as 7000 daltons and RKB and BS24 were inhibited by a 3200 dalton fraction. One, RKB, was inhibited by a fraction with a molecular weight of 2000. This fraction was further separated by preparative

HPLC on a 3 x 30 cm column of C_{18} Poresil B. The column was eluted with step gradients of acetonitrile (10%, 18%, 25%, 31%, 36%, 40% and 43%) in 50 mM ammonium formate buffer (pH 3.18). Eleven fractions were resolved by the column. Each fraction was lyophilized. From 2 to 5 mg of peptide was obtained from each major peak, which is the amount of material expected on the assumption that bovine Tg is a dimer. A fraction eluted with 36% acetonitrile, inhibited agglutination of bovine Tg-coated erythrocytes by hybridoma RKB. Sequence studies of this fraction are in progress.

SUMMARY

Using monoclonal antibodies we have begun to define the epitopes of the murine thyroglobulin molecule that elicit autoimmune responses. Based on the principle of complementation, 18 monoclonal antibodies were classified into 8 groups, defined by their reactions with the same or neighboring determinants. Further distinctions between the monoclonals were drawn from comparisons of their cross-reactions with thyroglobulins of other species and their patterns in isoelectric focusing. One low molecular weight fragment of bovine thyroglobulin, which cross-reacts extensively with thyroglobulins of other species, has been isolated and partially characterized.

ACKNOWLEDGEMENTS

This work was supported in part by National Institutes of Health grants AM 20023 and CA 18900.

REFERENCES

1. N. R. Rose, L. D. Bacon, R. S. Sundick, Y. M. Kong, P. S. Esquivel and P. E. Bigazzi, Genetic regulation in autoimmune thyroiditis, in: "Autoimmunity: Genetic, Immunologic, Virologic, and Clinical Aspects," N. Talal, ed., Academic Press, New York (1977).
2. A. O. Vladutiu, and N. R. Rose, Autoimmune murine thyroiditis: Relation to histocompatibility (H-2) type, Science 174:1137 (1971).
3. Y. M. Kong, C. S. David, A. A. Giraldo, M. ElRehewy and N. R. Rose, Fine structure of genetic control of autoimmune response to mouse thyroglobulin, in: "Genetic Control of Autoimmune Disease," N. R. Rose, P. E. Bigazzi and N. L. Warner, eds., Elsevier/North-Holland, New York (1978).
4. P. S. Esquivel, N. R. Rose and Y. M. Kong, Induction of autoimmunity in good and poor responder mice with mouse thyroglobulin and lipopolysaccharide, J. Exp. Med. 145:1250 (1977).

5. V. Tomazic and N. R. Rose, Autoimmune murine thyroiditis. VIII. Role of different thyroid antigens in the induction of experimental autoimmune thyroiditis, Immunology 30:63 (1976).
6. V. Tomazic, N. R. Rose, and D. C. Shreffler, Autoimmune murine thyroiditis. IV. Localization of genetic control of the immune response, J. Immunol. 112:965 (1974).
7. K. W. Beisel, C. S. David, A. A. Giraldo, Y. M. Kong, and N. R. Rose, Regulation of experimental autoimmune thyroiditis: Mapping of susceptibility to the I-A subregion of the mouse H-2, submitted (1981).
8. Y. M. Kong, C. S. David, A. A. Giraldo, M. ElRehewy and N. R. Rose, Regulation of autoimmune response to mouse thyroglobulin: Influence of H-2D-end genes, J. Immunol. 123:15 (1979).
9. N. R. Rose, Y. M. Kong, I. Okayasu, A. A. Giraldo, K. Beisel and R. Sundick, T-cell regulation in autoimmune thyroiditis, Immunol. Rev. 55:299 (1981).
10. I. M. Roitt, P. N. Campbell, and D. Doniach, The nature of the thyroid auto-antibodies present in patients with Hashimoto's thyroiditis (lymphadenoid goitre), Biochem. J. 69:248 (1958).
11. L. Nye, L. C. P. DeCarvalho, and I. M. Roitt, Restrictions in the response to autologous thyroglobulin in the human, Clin. exp. Immunol. 41:252 (1980).
12. H. Metzger, G. C. Sharp, and H. Edelhoch, The properties of thyroglobulin. IX. The molecular properties of iodinated thyroglobulin, Biochem. 1:205 (1962).
13. W. A. Stylos, and N. R. Rose, Splitting of human thyroglobulin. II. Enzymatic digestion, Clin. Exp. Immunol. 5:285 (1969).
14. W. A. Stylos, and N. R. Rose, Splitting of human thyroglobulin. IV. The antigenicity of the pepsin derived fragments, Clin. Exp. Immunol. 27:245 (1977).
15. H. Salabe, R. Dominici, G. B. Salabe, Immunological properties of Tg carbohydrates: enhancement of Tg immunoreaction by removal of sialic acid, Clin. Exp. Immunol. 25:234 (1976).
16. L. A. Nilsson, N. R. Rose, and E. Witebsky, Spontaneous thyroiditis in the obese strain of chickens. VI. Thyroxine-binding antibodies, J. Immunol. 107:997 (1971).
17. Y. Kondo, and Y. Kamiya, Some properties of the purified microsome-bound hog thyroid thyroglobulins: Possible intermediates in the maturation process of thyroglobulin, in: "Thyroid Research," J. Robbins and E. L. Braverman, eds. Academic Press, New York (1975).
18. M. J. Spiro, Studies on the protein portion of thyroglobulin. Amino acid composition and terminal amino acid of several thyroglobulins, J. Biol. Chem. 245:5820 (1970).
19. A. Vignal, B. M. Nataf, and M. Tubiana, Iodoamino acid composition of poorly iodinated human thyroglobulin fractionated by isopycnic centrifugation, Acta Endocrin. 87:320 (1978).
20. T. Arima, and R. G. Spiro, Studies on the carbohydrate units of thyroglobulin. Structure of the mannose-n-acetylglucosamine

unit (unit A) of the human and calf proteins, J. Biol. Chem. 247:1836 (1972).
21. T. Kondo, M. Fukuda, and T. Osawa, The structure of unit B-type glycopeptides from porcine thyroglobulin, Carbohyd. Res. 58:405 (1977).
22. M. J. Spiro, and R. G. Spiro, Lipid-saccharide intermediates in thyroid glycoprotein synthesis, in: "Thyroid Research," J. Robbins and E. L. Braverman, eds., Academic Press, New York (1975).
23. A. L. Tarentino, T. H. Plummer, and F. Maley, A β-mannosidic linkage in the unit A oligosaccharide of bovine thyroglobulin, J. Biol. Chem. 248:5547 (1973).
24. O. Tarutani, and S. Shulman, Properties of carbohydrate-stripped thyroglobulin. I. Preparation and physicochemical properties of desialized thyroglobulin, B.B.A. 229:642 (1971).
25. O. Tarutani, and S. Shulman, Properties of carbohydrate-stripped thyroglobulin. II. Heterogeneity in sialic acid content in thyroglobulin subfractions, B.B.A. 236:384 (1971).
26. M. Z. Atassi, Antigenic structure of myoglobin: The complete immunochemical anatomy of a protein and conclusions relating to antigenic structures of proteins, Immunochem. 12:423 (1975).
27. M. Z. Atassi, Precise determinations of the entire antigenic structure of lysozyme: molecular features of protein antigenic structures and potential of 'surface-simulation' synthesis - a powerful new concept for protein binding sites, Immunochem. 15:909 (1978).
28. L. Adorini, M. Harvey, and E. E. Sercarz, The fine specificity of regulatory T cells. IV. Idiotypic complementarity and antigen-bridging interactions in anti-lysozyme response, Eur. J. Immunol. 9:906 (1979).
29. L. Adorini, M. Harvey, A. Miller, and E. E. Sercarz, Fine specificity of regulatory T cells. II. Suppressor and helper T cells are induced by different regions of hen egg-white lysozyme in a genetically nonresponder mouse strain, J. Exp. Med. 150:293 (1979).
30. G. A. Hashim, and E. H. Eylar, Allergic encephalomyelitis: Isolation and characterization of encephalitogenic peptides from the basic protein of bovine spinal cord, Arch. Biochem. Biophys. 129:645 (1969).
31. C. -H. J. Chou, F. C. -H. Chou, T. J. Kowalski, R. Shapira, and R. F. Kibler, The major site of guinea pig myelin basic protein encephalitogenic in Lewis rats, J. Neurochem. 28:115 (1977).
32. J. E. Swierkosz, and R. H. Swanborg, Immunoregulation of experimental allergic encephalomyelitis: Conditions for induction of suppressor cells and analysis of mechanism, J. Immunol. 119:1501 (1977).
33. A. J. Crowle, "Immunodiffusion," Academic Press, New York (1961).

34. R. B. Goudie, J. R. Anderson, and K. G. Gray, Non-precipitating anti-thyroglobulin studied by the Ouchterlony technique, Immunol. 2:300 (1959).
35. N. R. Rose, J. H. Kite, Jr., T. K. Doebbler and E. Witebsky, Chronic thyroiditis as an autoimmune disease, in: "Injury, Inflammation and Immunity," L. Thomas, ed., Williams & Wilkins, Baltimore (1964).
36. J. Steensgaard, C. Jacobsen, J. Lowe, D. Hardie, N. R. Ling and R. Jefferis, The development of difference turbidimetric analysis for monoclonal antibodies to human IgG, Mol. Immunol. 17:1315 (1980).
37. Y. M. Kong, N. R. Rose, M. ElRehewy, R. Michaels, A. A. Giraldo, M. A. Accavitti and M. A. Leon, Thyroid alloantigens in autoimmunity. Trans. Proc. 12:129 (1980).

3

THE MAJOR B AND T CELL DETERMINANT ON PIGEON CYTOCHROME *c* IN B10.A MICE

Charles Hannum[1], Michiel Ultee[1], Louis A. Matis[2], Ronald H. Schwartz[2], and E. Margoliash[1]

[1]Department of Biochemistry, Molecular, and Cell Biology Northwestern University, Evanston, IL 60201
[2]The Laboratory of Immunology, National Institute of Allergy and Infectious Diseases, National Institutes of Health, Bethesda, MD 20205

Abstract

The specificities of B10.A B and T cells responding to pigeon cytochrome *c* have been examined. Proliferating T cells recognize glutamine 100 and lysine 104 and can be stimulated *in vitro* by either native cytochrome *c* molecules or certain of their CNBr-cleavage fragments. In contrast only molecules with the native cytochrome *c* conformation were found to interact with B10.A anti-pigeon cytochrome *c* antibodies. Antibodies appear to recognize a determinant or determinants which overlaps with that which elicits the T cell proliferative response.

B CELL RESPONSES TO VARIOUS CYTOCHROMES *c*

Analyses of the specificities of antibodies raised against protein antigens have generally been limited by the lack of sets of cross-reactive molecules sufficient for the examination of a meaningful fraction of the possible structures of the antigenic determinants. To circumvent this limitation, fragments of the antigen, obtained by cleavage or synthesis, are tested either for their abilities to interact directly with the antibodies or to inhibit the binding of the antibodies to the native antigen.[1] Although the affinities of antibodies raised against the intact protein are sometimes orders of magnitude lower for the fragments than for the original molecule[2], this approach has been used to locate antigenic determinants in many cases, such as myoglobin[3] lysozyme,[4] bovine serum albumin,[5] tobacco mosaic virus,[6] ferredoxin[7] and staphylococcal nuclease[8]. The question remains, however, whether certain topographic determinants cannot be found by such fragments either because the proportion of the fragments remaining in the conformations they had in the native protein is too small to be detected or because the original topographic determinants consisted of segments widely separated in the polypeptide chain but brought into proximity by the spatial arrangement of the protein.

Cytochrome c provides perhaps the most extensive set of cross-reactive protein molecules so far available, with over 80 eukaryotic cytochromes c having been sequenced[9,10]. To this list can be added many chemically modified forms of the protein[11,12] as well as a technique for synthesizing hybrid native molecules from the CNBr-cleavage fragments of virtually any two cytochromes c[13-16]. The most compelling advantage of this protein as a tool for mapping antigenic determinants is the finding that all cytochromes c have seemingly identical spatial conformations and differ from one another at only a few side chain substitutions[17-23]. Thus, native cytochromes c provide a valuable set of molecules for determining the contribution of topographic alterations to immunogenicity and receptor specificity.

In previous work, we have analyzed the humoral immune responses of rabbits to human[24-26], guanaco[27,28], mouse[27,28], horse[29] and rabbit[30] cytochromes c, and of mice to guanaco and rabbit cytochromes c[27,28]. The general approach used has involved the passage of the antisera through series of affinity columns containing, as ligands, cytochromes c which share only certain of the sequence differences between the rabbit protein and the immunogen[31]. This leads to the separation of individual populations of antibodies directed toward single sites on the surface of the molecule. To demonstrate that these are indeed single site populations, the stoichiometries of interaction of the corresponding Fab fragments with the immunogen are determined by titrating the quenching of antibody-tryptophan fluorescence on binding with the heme-containing antigen[32]. To further analyze the influence of each amino acid residue involved in each determinant, fluorescence quenching titrations and competitive binding assays are carried out with cytochromes c having structures appropriately related to that of the immunogen.

It was found that generally antibodies are produced only against surface amino acid sequence differences between the immunogen and the homologous host cytochromes c. With persistent boosting, however, rabbits do respond, albeit very weakly, to their own cytochrome c with antibodies directed against those portions of the molecule which are, on an evolutionary time-scale, the most variable among mammalian cytochromes c, namely the regions on the surface of the molecule which include residues 44, 62 and 89[30]. It would appear that there are selective pressures which act during evolution to eliminate, from the antibody repertoire, specificities for self-antigenic determinants. One would expect to detect such self-specificities corresponding to the areas which have varied most recently in the course of the protein's evolution and which, therefore, have not had enough time to be eliminated by the slow selective process. This is presumably the situation in the case of the rabbit response to rabbit cytochrome c.

An overall summary of the results with mammalian cytochromes c as immunogens is that rabbits produce antibody populations against the regions surrounding and including residues 44, 62, and 89 on mouse and probably on rabbit cytochromes c, residues 62 and 89 on guanaco cytochrome c, residues 44, 60, 89 and 92 on horse cytochrome c, and, residue 58 among others, on human cytochrome c. A/J mice produce antibody populations against the regions surrounding and including residues 44 and 62 on guanaco cytochrome c and residues 44, 62 and 89 on rabbit cytochrome c.

The binding of an antibody population directed toward a region containing a variant amino acid is affected not only by that residue, but also by other residues in the immediate vicinity. For example, one sub-population of rabbit-anti-horse cytochrome c antibodies was shown, using competitive binding assays, to be affected by residues 88, 89, 92 and 96, all of which are on the surface of the protein, clustered within 12 Å of one another on the carboxyl-terminal α-helix[29]. Such antibodies will specifically bind to those regions on the immunogen or any other cytochromes c having the identical or appropriately similar structures in those areas. Additionally, the binding of one antibody to one determinant is totally independent of the binding of other antibodies to distal determinants on the molecule[28]. Since B cell receptors have binding characteristics which are identical to those of the corresponding antibody[33], these and similar results obtained in other laboratories paint a fairly good picture of the manner in which B cells bind antigenic determinants on globular proteins.

In summary, B cells responding to a protein immunogen with a very stable tertiary structure, such as cytochrome c, are affected by antigenic determinants which act by virture of the contours on the surface of the immunogen and their physico-chemical properties insofar as they differ from the corresponding areas on the host's homologous protein. However, the receptors on the responding B cells clearly bind more than just the side chains which differ, including the surrounding familiar terrain.

THE PROLIFERATIVE RESPONSE OF T CELLS
FROM B10.A MICE TO PIGEON CYTOCHROME c

The T cell response to cytochrome c was analyzed employing inbred mice and the PETLES proliferation assay. PETLES is an acronym for peritoneal exudate T lymphocyte-enriched subpopulations. This assay has been shown, with other antigens such as GLØ, to be sensitive to Ir-gene control. Inbred strains, as well as recombinant and congenic-resistent lines of mice, enable the mapping of the genetic control of immune responses to subregions within the MHC (Major Histocompatibility Complex). Moreover, inbred mice often produce fairly restricted responses to simple antigens. This was found to be the case with cytochromes c which differ at only a few residues from the homologous mouse protein. Thus one can ask whether B and T cells respond to the same or different determinants on a globular protein antigen. Indeed, such a relatively simple system should make it straightforward to sort out which responses result from which reactions with which antigenic determinants.

With these considerations in mind, a survey was conducted[34]. Numerous mouse strains and cytochromes c as immunogens were tested to find a response under Ir-gene control that can be examined for specificity by cross-reactivities with other readily available cytochromes c. Also, the response should exhibit very little auto-immune nature such that little or no cross-reactivity exist with mouse cytochrome c itself, making it possible to attribute any observed cross-reactivities to the amino acid residues which differ between the immunogen and the host cytochrome c. The system ultimately selected, the B10.A mouse response to pigeon cytochrome c, met these requirements, in that the cytochrome c was highly immunogenic in this strain, the resultant PETLES showed no stimulation with mouse cytochrome c, and a large set of cytochromes c of known amino acid sequence, was available for the investigation of each significant sequence position. PETLES from animals primed with native pigeon cytochrome c were found to proliferate in vitro following stimulation with either native or fragmented forms of the molecule.

The PETLES response to pigeon cytochrome c was found to be controlled by at least two MHC-linked Ir genes, one mapping to I-A and the other to I-E[35]. A similar situation has been found with GLØ[36] in which response apparently requires the generation of a hybrid Ia antigen resulting from the interaction of Ir-gene products from both I-subregions[37,38].

From comparative studies with different cytochromes c and their CNBr-cleavage fragments, it was found that full stimulation of native pigeon cytochrome c-primed PETLES requires an uncharged residue at position 100 (at which pigeon carries a glutamine) and a lysine at the carboxyl-terminus (residues 103 or 104). Additionally, isoleucine 3, which is adjacent to glutamine 100 in the tertiary structure, was also implicated as possibly being involved in stimulation[39]. Experiments in which cytochromes c were mixed which bear only one or the other of the two required residues, namely glutamine 100 and lysine 104, showed no additivity in stimulation, indicating that the receptors interacting with this antigen recognize glutamine 100 and lysine 104 together. In all but one instance, PETLES primed to native pigeon cytochrome c were more strongly stimulated by the carboxyl-terminal CNBr fragments, consisting of either residues 81-103 or 81-104, than by the native cytochromes which stimulated. For example, stimulation by duck cytochrome c fragment 81-104 exceeds that seen with intact duck cytochrome c.

Figure 1 is a molar dose-response curve showing that the carboxyl-terminal pigeon fragment 81-104, native tobacco hornworm moth cytochrome c, and moth 81-103 are all more stimulatory than native pigeon cytochrome c, the original immunogen, with moth 81-103 resulting in a 100-fold shift in the curve. This hetero-

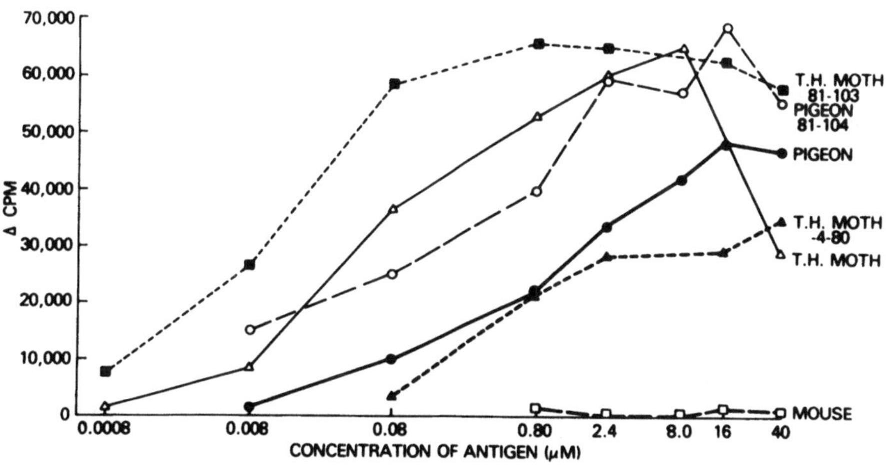

Figure 1. The T lymphocyte proliferative responses to pigeon (●—●), mouse (□--□) and tobacco hornworm (T.H.) moth (△—△) cytochromes c and to the peptide fragments 81-104 of pigeon cytochrome c (○--○), -4 to 80 of T.H. moth cytochrome c (▲--▲) and 81-103 of T. H. moth cytochrome c (■--■). The PETLES were from B10.A mice immunized with 1.6 nmoles of pigeon cytochrome c. The abscissa represents the concentrations of cytochromes c added to the in vitro PETLES cultures, and the ΔCPM on the ordinate represents the radioactivity accumulated by the cells following a 1μCi pulse of tritiated thymidine 16 to 18 hours prior to harvesting.

cliticity is also seen with T cell clones[40] and suggests that the receptors governing this response have a higher affinity for the determinant if the terminal lysine is at position 103 rather than 104. The increased stimulation seen with the CNBr fragment 81-103 of tobacco hornworm moth cytochrome c may result from an increased flexibility of the peptide allowing the lysine 103 to assume the conformation required for optimal stimulation.

T cells appear to exhibit limited stringency with respect to antigen conformation since native cytochrome c-primed PETLES are highly responsive to fragments of the immunogen. However, it should be noted that glutamine 100 and lysine 104 are part of an α-helix which sweeps down across the top back surface of the native molecule, and it is possible that a significant fraction of the peptides extending from 81-104 may retain, in solution, the α-helical conformation.

B10.A B CELL RESPONSES TO PIGEON CYTOCHROME c

Because mice cannot readily provide antisera in the quantities that were earlier used in the analyses of the rabbit responses to human, guanaco, mouse and rabbit cytochromes c, examination of the B10.A B cell response to pigeon cytochrome c has proven to be somewhat more difficult. As discussed above, the earlier studies depended on the separation of single-site antibody populations by specific adsorption on affinity columns of various cytochromes c. In the present case we developed the capability of detecting different antibody populations in the unfractionated serum. Pairs of cytochromes c were either found or constructed which differed from one another at only one portion of the molecule, one molecule looking like pigeon cytochrome c at the tested position and the other not. If only one of the two cytochromes could inhibit the interaction of pigeon cytochrome c with the antibodies present in the serum, then it became clear that that residue or region of difference between the two competitors is of significance in the B cell recognition of pigeon cytochrome c

The assay used is an adaptation of that developed by Klinman[41] for the detection of anti-cytochrome c antibodies produced by splenic foci[42]. Briefly, antibodies are bound to the walls of microtitering wells coated with the immunogen, in this case pigeon cytochrome c. If a competing cytochrome c is added to the wells prior to the addition of the antiserum, the inhibition of antibody binding will be a measure of that cytochrome's ability to react with antibodies directed against the cytochrome c on the wall of the well. A comparison of competitor sequences allows a mapping of the site responsible for the competition.

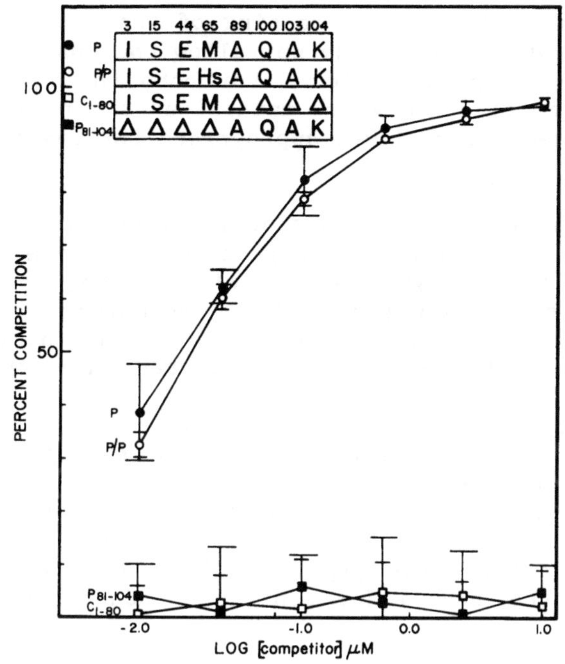

Figure 2. Competitive binding assay in which P, pigeon cytochrome c (●—●), P/P, pigeon cytochrome c reconstituted from its 1-65 and 66-104 CNBr-cleavage fragments (○—○), P_{81-104}, the carboxyl-terminal CNBr fragment of pigeon cytochrome c (■—■), and C_{1-80}, the amino-terminal CNBr fragment of chicken cytochrome c which is identical to an equivalent peptide of pigeon cytochrome c (□—□), are compared for their abilities to inhibit the binding of B10.A anti-native pigeon cytochrome c Fab fragments to microtitering wells coated with the immunogen. The error bars represent the 95% confidence level by the student's t-test. The key at the top identifies each competitor and lists the amino acid residues each carries at the seven positions at which pigeon cytochrome c differs from mouse cytochrome c and also residue 65, the point at which the two fragments are fused in the preparation of the hybrid molecules. The single-letter code for the amino acids is used, with Hs designating homoserine and Δ representing a residue deletion.

Figure 2 shows that fragments are of little value in this analysis since fragments of the original immunogen, pigeon cytochrome c, show little or no ability to inhibit the binding of B10.A anti-native pigeon cytochrome c Fab fragments to the native protein, even at concentrations two and one half orders of magnitude higher than that required for 50% inhibition by the native immunogen. Furthermore, the available native cytochromes c, although very adequate for the mapping of the T cell determinant through the use of fragments, do not allow the unambiguous elucidation of antibody specificities. For both these reasons cytochromes c were constructed which maintained the native conformation and contained the residues appropriate for testing antigenic structure.

Since lysine 104 is a residue of interest, molecules of des-lysyl 104-pigeon cytochrome c (Pigeon 1-103) were produced by treating native pigeon cytochrome c with carboxypeptidase B for one minute and separating the products by column chromatography on carboxymethylcellulose. The removal of a full positive charge allows the product to be cleanly separated from the reactant.

The technique for the production of hybrid cytochrome c molecules from CNBr fragments of cytochromes from different species, which was developed by Corradin and Harbury[13-16], was modified. The carboxyl-terminal homoserine of the 1-65 fragment was fully cyclized to the lactone by treatment with anhydrous trifluoroacetic acid. This peptide, which is more readily obtained than is the 66-104 fragment, was added in large molar

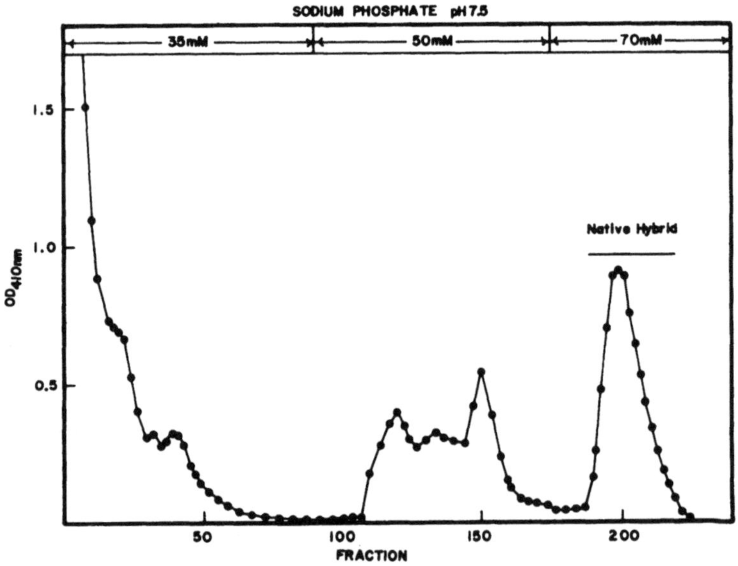

Figure 3. Elution profile of the purification of a reconstituted pigeon cytochrome c by cation exchange chromatography on carboxymethylcellulose. The bar above the final peak designates the fractions pooled in the collection of the native reconstituted molecule. The top line represents the step-gradient employed in obtaining the chromatogram.

excess to the reaction mixture, and the product was purified to homogeneity by cation exchange chromatography on carboxymethylcellulose. Figure 3 is the elution profile of such a chromatogram, in which a pigeon cytochrome c, reconstituted from its fragments, was purified from the remainder of the reaction mixture. The peak marked "native hybrid" contains pigeon cytochrome c differing from the original material only by having a homoserine, rather than a methionine, at residue 65. A large variety of such molecules has now been constructed.

Since such hybrid cytochromes c are available in limited quantities, only a relatively low range of competitor concentrations could be spanned in the assays. To circumvent this problem the assay sensitivity was increased by producing monovalent Fab fragments from the antibodies to be examined. In this fashion, at competitor concentrations which would show little effect with whole antisera, 90% competitions can be obtained. It would appear that the binding affinity of a bivalent antibody to an insolubilized antigen is sufficiently higher than that of an Fab fragment that much higher competitor concentrations are required to displace them.

To determine whether the construction of hybrid cytochromes c has any effect on the ability of antibodies raised against native pigeon cytochrome c to bind to a reconstituted molecule, parallel competition assays were carried out with the two forms of pigeon cytochrome c. In Figure 2, the curves for the native and hybrid forms of the molecule are indistinguishable, showing that the hybridization procedure, in itself, does not result in a product with substantially different affinity for its specific antibodies.

In Figure 4, P/P, a reconstituted pigeon cytochrome c, competes to 50% at $10^{-1.6}$M cytochrome c, whereas mouse cytochrome c requires a 290-fold higher concentration to reach this point. The only differences between these two molecules are the seven side-chain substitutions illustrated in Figure 5. The binding energy of any receptor-ligand interaction is defined by the equation $\Delta G = -RT \ln K_a$, where K_a is the association constant From the data presented in Figure 4, it is possible to roughly calculate the contribution of the side-chain substitutions to the overall binding energy of B10.A anti-pigeon cytochrome c Fab fragments with the pigeon protein, assuming an average K_a. The side-chain difference contribution would be a $\Delta\Delta G$ calculated from the following:

$$\Delta\Delta G = (-RT \ln K)_{pigeon} - (-RT \ln K)_{mouse}$$

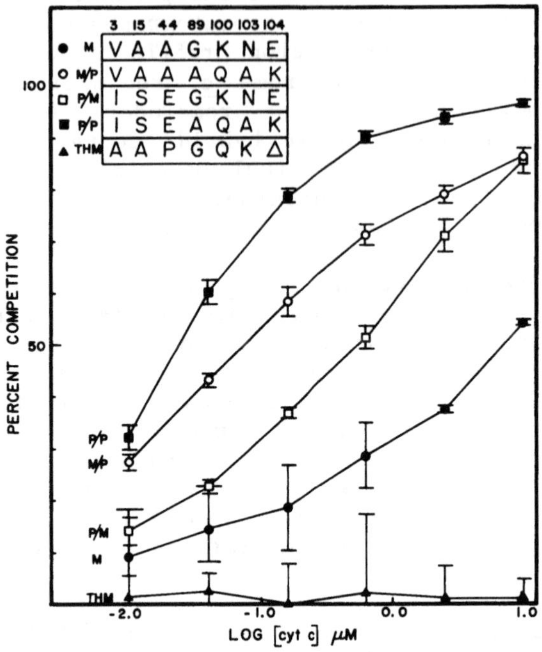

Figure 4. Competitive binding assay as given in the legend to Figure 2, in which P/P, a reconstituted pigeon cytochrome c (■—■), M/P, a hybrid mouse/pigeon cytochrome c constructed from the 1-65 CNBr fragment of mouse cytochrome c and the 66-104 CNBr fragment of pigeon cytochrome c (○—○),P/M, a hybrid pigeon/mouse cytochrome c (□—□), M, mouse cytochrome c (●—●), and THM, tobacco hornworm moth cytochrome c (▲—▲) are compared for their abilities to prevent the binding of B10.A anti-native pigeon cytochrome c Fab fragments to the walls of microtitering wells coated with the immunogen.

Since the comparison is made at 50% antibody bound and the same amount of antibody was added to each reaction, the equation simplifies to:

$$\Delta\Delta G = -RT \ln \frac{[\text{cytochrome } \underline{c}]_{\text{mouse}}}{[\text{cytochrome } \underline{c}]_{\text{pigeon}}}$$

In the present case we obtained a value of $\Delta\Delta G = -3.3$kcal/mole, this constituting the contribution of the side-chain differences between mouse and pigeon cytochromes c to the free energy of interaction with anti-pigeon cytochrome c Fab fragments. If one varies K_a over as wide a range as $10^5 M^{-1}$ to $10^{10} M^{-1}$, the total binding energy varies from -8.2 to -14.9 kcal/mole, and the $\Delta\Delta G$

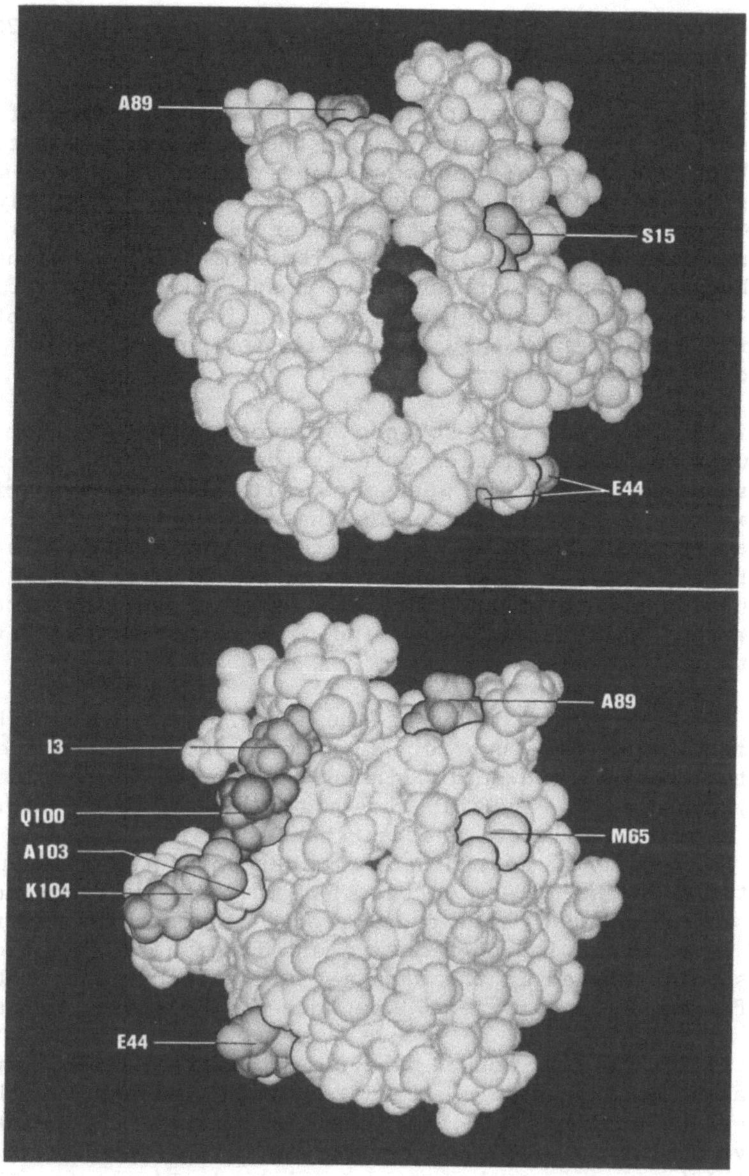

Figure 5. *Front (top) and back (bottom) views of a computer-generated space-filling model of pigeon cytochrome c. The seven positions at which pigeon cytochrome c differs from the mouse protein are indicated by lines following the contours of the atoms representing each residue. Residue 65, the site of peptide fusion during hybrid synthesis is similarly identified. The authors are grateful to Mr. Richard Feldmann, of the Division of Computer Research and Technology, National Institutes of Health, for the computer graphics model.*

for the variant amino acid side-chains constitutes from 40.2% to 22.1% of the total.

This clearly indicates that in any case the binding of the protein surface contours surrounding the side-chain substitutions, which are the same on mouse and pigeon cytochromes c, must provide the majority of the binding energy. This may be why a weak autoimmune response is observed in rabbits to repeated immunizations with rabbit cytochrome c[30]. Although the lymphocyte receptors present have comparatively low affinities for the self-molecule, high concentrations of this protein over a prolonged period may eventually trigger some of the clones to respond.

It is of interest that tobacco hornworm moth cytochrome c, which causes a heteroclitic stimulation of pigeon-specific PETLES, shows no interaction whatsoever with the pigeon-specific Fab fragments. This fact indicates that either the determinant recognized by the T cells is very different and perhaps distant on the protein surface from that recognized by the B cells, or that even if the determinants are the same or overlapping then the parameters which affect the binding must be very different for the two cell types.

The relative competitions of the mouse/pigeon (M/P) hybrid cytochrome c, that is a hybrid composed of the 1-65 peptide of mouse cytochrome c covalently linked to the 66-104 peptide of pigeon cytochrome c, and of the pigeon/mouse (P/M) hybrid, shown in Figure 4, indicate that antibodies against pigeon cytochrome c recognize a determinant or determinants created by both portions of the molecule, but that most of the reaction is with the 66-104 segment. The cross-reactivity patterns observed at the T cell level with both native and fragmented cytochromes c are also distributed in the same manner, although the contribution of the 1-65 fragment was not as clearly defined.

To illustrate the method whereby potential sites can be specifically implicated or eliminated with respect to their involvement in antibody binding, one can compare the relative cometitive abilities of pigeon cytochrome c, the immunogen, and des-lysyl 104 pigeon cytochrome c, as shown in Figure 6. These two molecules differ only at lysine 104, and it is quite apparent that the loss of this residue has greatly affected antibody binding. From calculations identical to those described above, it would appear that this residue alone can account for about 20% of the variant side-chain contribution to the binding energy. In experiments with antibodies from A/J mice immunized with native pigeon cytochrome c, des-lysyl 104 pigeon cytochrome c showed competition equal to that of the native molecule, indicating that the loss of the carboxyl-terminal lysine has

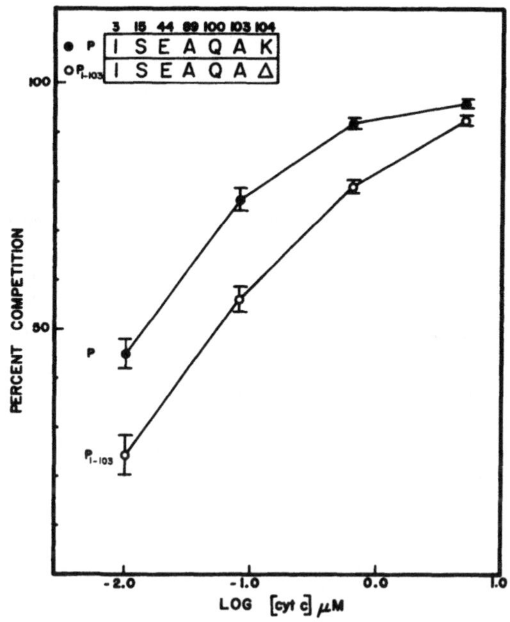

Figure 6. Competitive binding assay given in the legend to Figure 2, in which P, pigeon cytochrome c (●—●), and P_{1-103}, des-lysyl-104 pigeon cytochrome c (○—○) are compared for their abilities to block the binding of B10.A anti-native pigeon cytochrome c Fab fragments to microtitering wells coated with the immunogen.

not otherwise perturbed the antigenic structure of the protein.

In summary, these preliminary results show that the B10.A B cell response to pigeon cytochrome c is directed towards a determinant or determinants generated by amino acid residues on both the 1-65 and 66-104 fragments of the molecule, and that, at least to the limits to which the assay has thus far been extended, it appears that such determinants only exist when the protein is in its native form. As has also been seen in the T cell response, most of the contribution to antibody binding is associated with the carboxyl-terminal portion of the protein, although the amino terminal segment also provides an important contribution to the B cell recognition. This last point remains unclear with respect to the T cells. Lysine 104 has definitely been implicated in B cell recognition, but it is only providing a fraction of the binding energy. Whether lysine 104 is a determinant in and of itself or is part of a complex determinant involving other residue substitutions remains to be resolved. The present data does, however, show that at least a portion of the B cell response is directed against the same region of the molecule as is recognized by the T cells. A more thorough analysis is in progress.

REFERENCES

1. Reichlin, M. (1975) Adv. Immunol. 20: 71.
2. Berzofsky, J. A., Schechter, A, N., Shearer, G. M. and Sachs, D. H. (1977) J. Exp. Med. 145: 123.
3. Atassi, M. Z. and Kazim, A. L. (1978) in Immunobiology of Proteins and Peptides-I. Edited by Atassi, H. Z. and Stavitsky, A. B. Plenum, New York. Adv. Exp. Med. Biol. 98: 19.
4. Atassi, M. Z. (1978) in Immunobiology of Proteins and Peptides-I. Edited by Atassi, H. Z. and Stavitsky, A. B. Plenum, New York. Adv. Exp. Med. Biol. 98: 41.
5. Habeeb, A. F. S. A. (1978) in Immunobiology of Proteins and Peptides-I. Edited by Atassi, H. Z. and Stavitsky, A. B. Plenum, New York. Adv. Exp. Med. Biol. 98: 101.
6. Benjamini, E., Leung, C. Y., and Rennick, D. M. (1978) Immunobiology of Proteins and Peptides-I. Edited by Atassi, H. Z. and Stavitsky, A. B. Plenum, New York. Adv. Exp. Med. Biol. 98: 165.
7. Kelly, B. and Levy, J. G. (1978) in Immunobiology of Proteins and Peptides-I. Edited by Atassi, H. Z. and Stavitsky, A. B. Plenum, New York. Adv. Exp. Med. Biol. 98: 181.
8. Berzofsky, J. A., Pisetsky, D. S., Schwartz, R. H., Schechter, A. N., and Sachs, D. H. (1978) in Immunobiology of Proteins and Peptides-I. Edited by Atassi, H. Z. and Stavitsky, A. B. Plenum, New York. Adv. Exp. Med. Biol. 98: 241.
9. Dayhoff, M. O., and Eck, R. V. (1972) Atlas of Protein Sequence and Structure, Vol. 5, Suppl. 2, National Biochemical Research Foundation, Washington, D. C.
10. Borden, D., and Margoliash, E. (1976) in Handbook of Biochemistry and Molecular Biology (Fasman, G. D., ed.) pp. 268-279, The Chemical Rubber Co., Cleveland.
11. Ferguson-Miller, S., Brautigan, D. L., and Margoliash, E. (1979) in The Porphyrins (Edited by Dolphin, D.). Academic Press, New York. 7: 170-181.
12. Brautigan, D. L., Ferguson-Miller, S., and Margoliash, E. (1978) J. Biol. Chem. 253: 130.
13. Corradin, G. and Harbury, H. A. (1970) Biochem. Biophys. Acta 221: 489.
14. Corradin, G. and Harbury, H. A. (1971) Proc. Natl. Acad. Sci. U.S.A. 68: 3036.
15. Corradin, G. and Harbury, H. A. (1974) Biochem. Biophys. Res. Commun. 61: 1400.
16. Corradin, G. and Harbury, H. A. (1974) Fed. Proc. 33: 1302.
17. Margoliash, E. (1972) Harvey Lect. 66: 177.
18. Dickerson, R. E. and Timkovich, R. (1975) Cytochromes c. In The Enzymes, 3rd. Ed. Vol. 11. Edited by P. D. Boyer. Academic Press, New York, p. 397.

19. Dickerson, R. E., Takano, T., Eisenberg, D., Kallai, O. B., Samson, L., Cooper, A., and Margoliash, E. (1971) J. Biol. Chem. 246:1511.
20. Salemme, F. R., Freer, S. T., Yuong, N. H., Alden, R. A., and Kraut, J. (1973) J. Biol. Chem. 248: 3910.
21. Salemme, F. R., Kraut, J., and Kamen, M. D. (1973) J. Biol. Chem. 248: 7701.
22. Timkovich, R. and Dickerson, R. E. (1973) J. Mol. Biol. 79: 39.
23. Timkovich, R. and Dickerson, R. E. (1976) J. Biol. Chem. 251: 4033.
24. Nisonoff, A., Reichlin, M. and Margoliash, E. (1970) J. Biol. Chem. 254: 940.
25. Noble, R. W. Reichlin, M., and Schreiber, R. D. (1972) Biochemistry 11: 3326.
26. Smith, L., Davies, H. C., Reichlin, M. and Margoliash, E. (1973) J. Biol. Chem. 248: 237.
27. Urbanski, G. and Margoliash, E. (1976) in The Immunochemistry of Enzymes and their Antibodies. Edited by M. Salton. J. Wiley and Sons, New York, p. 203.
28. Urbanski, G. and Margoliash, E. (1977) J. Immunol. 118: 1170.
29. Jemmerson, R. and Margoliash, E. (1979) J. Biol. Chem. 254: 12706.
30. Jemmerson, R. and Margoliash, E. (1979) Nature 282: 468.
31. Jemmerson, R. and Margoliash, E. (1981) Methods Enzymol. 74: 244.
32. Nobel, R. W., Reichlin, M., and Gibson, Q. H. (1969) J. Biol. Chem. 244: 2403.
33. Cosenza, H. and Kohler, H. (1972) Proc. Natl. Acad. Sci. U.S.A. 69: 2701.
34. Schwartz, R. H., Solinger, A. M., Ultee, M., and Margoliash, E. (1978) in Immunobiology of Proteins and Peptides - I. Edited by M. Z. Atassi and A. B. Stavitsky. Plenum, New York. Adv. Exp. Med. Biol. 98: 371.
35. Solinger, A. M., Ultee, M. E., Margoliash, E., and Schwartz, R. H. (1979) J. Exp. Med. 150: 830.
36. Schwartz, R. H., Dorf, M. E., Benacerraf, B. and Paul, W. E. (1976) J. Exp. Med. 143: 897.
37. Schwartz, R. H., Yano, A., Stimpfing, J. H., and Paul, W. E. (1979) J. Exp. Med. 149: 40.
38. Jones, P. P., Murphy, D. B., and McDevitt, H. O. (1978) J. Exp. Med. 148: 925.
39. Ultee, M. E., Margoliash, E., Lipkowski, A., Flouret, G., Solinger, A. M., Lebwohl, D., Matis, L. A., Chen, C., and Schwartz, R. H. (1980) Molec. Immunol. 17: 809.
40. Matis, L. A. and Schwartz, R. H., personal communication.
41. Klinman, N. R. and Press, J. L. (1975) Transplant. Rev. 24: 49.
42. Speck, N. A. and Pierce, S. K., personal communication.

4

HELPER T CELLS SPECIFIC FOR PROTEIN ANTIGENS: ROLE OF SELF MAJOR HISTOCOMPATIBILITY COMPLEX AND IMMUNOGLOBULIN GENE PRODUCTS

Charles A. Janeway, Jr.,* Kim Bottomly,* Barry Jones, Patricia P. Jones, Ethan A. Lerner, Louis A. Matis, Janet M. McNicholas, Donal B. Murphy, and Ronald H. Schwartz

Department of Pathology, Yale University School of Medicine, New Haven, CT 06511
Department of Biological Sciences, Stanford University, Stanford, CA 94305
Laboratory of Immunology, NIAID, NIH, Bethesda, MD 20205

*Investigator, Howard Hughes Medical Institute

INTRODUCTION

Antibody responses to most antigens are dependent on the cooperative interaction of two distinct populations of lymphocytes. B lymphocytes are the precursors to antibody forming cells; their ability to produce antibody is dependent on helper T lymphocytes (Th). In classical studies of helper T cell function, B cells specific for haptens have been stimulated with hapten-protein conjugates, and the role of protein carrier-primed helper T cells measured (1). In such studies, the hapten must be physically linked to the carrier protein (1,2), and the helper T cell specifically primed to the carrier used for secondary immunization (1,2). Using this system, Katz and coworkers (3) made the surprising discovery that Th and B lymphocytes needed to be identical in the I region of the major histocompatibility complex (MHC) in order to cooperate effectively. Subsequent studies by a number of investigators have demonstrated that Th cells are selected during their development for the ability to recognize particular MHC encoded antigens as self, and that such Th cells will not respond to antigen except in the context of the MHC encoded antigen for which they are specific (4). This phenomenon has been termed MHC restriction.

The function of MHC restricted Th cells is intimately related to the action of specific immune response (Ir) genes that also map to the I region of the MHC. Katz and coworkers demonstrated that Th from responder by non-responder F1 mice would help B cells from the responder parent, but not those of the non-responder parent, in the response to DNP coupled to the genetically controlled antigen (5). Thus, the products of MHC genes, including the Ir genes, can influence the behavior of protein antigen specific helper T cells.

In studying the response to the hapten 2,4-dinitrophenyl (DNP), it was found that two distinct types of helper T cells act synergistically early in the adoptive secondary antibody response (2). Subsequent studies have shown that one of these helper T cells is carrier-specific, requires a hapten-carrier bridge, is absolutely required for *in vitro* anti-DNP antibody responses, and is most likely MHC restricted (6). The other helper T cell is also carrier-specific, but does not require a hapten-carrier bridge to act on B cells; it is not found in agammaglobulinemic mice, and it has therefore been proposed that such cells are in fact specific for self immunoglobulin gene products (7). This has been shown more directly by Bottomly et al. (8,9) in the antibody response to the hapten phosphorylcholine (PC), in which a second type of helper T cell has clearly been shown to act selectively on B cells bearing a particular idiotypic determinant.

The present chapter will examine the influence of MHC and Ig gene products on helper T cell function. The following points will be made, and supported with experimental data: First, that cloned, antigen-specific, MHC restricted Th cells can be prepared, and that their interaction with B lymphocytes is MHC restricted. Second, that reactivity to non-self MHC encoded antigenic determinants is a function of the same cells that respond to protein antigens in the context of self MHC encoded antigenic determinants. Third, that such MHC restricted cloned T cells do not have the Ig-dependent helper T cell that is required for early help in the anti-DNP antibody response or for dominant idiotype production in the anti-PC antibody response (10,11). Fourth, that immune response gene products are the Ia antigens recognized by the MHC restricted Th cells. And fifth, that the amount of Ir gene product expressed on a cell can greatly influence its functional activity. From these studies, two classes of protein antigen specific, synergistic helper T cells can be discriminated, one interacting with the B cell via Ia antigens, and the other via self immunoglobulin molecules.

MATERIALS AND METHODS

The materials and method used in these studies have all been described in detail in other publications; they will be referred to in the text as the various experimental systems are described.

RESULTS AND DISCUSSION

Cloned, Ovalbumin-Specific T Cells are MHC Restricted in the Proliferative Response to Antigen

We have prepared a number of ovalbumin-specific T cell lines cloned from primary culture in soft agar by the technique of Sredni et al. (2). Such T cell lines will grow indefinitely *in vitro* if stimulated with antigen, syngeneic feeder cells and T cell growth factor (TCGF). They maintain their specificity and function over prolonged periods of time. The responses of two such clones are given in Table 1. It can be seen that the clones respond specifically to OVA presented by syngeneic spleen cells, but do not respond to allogeneic stimulator cells in the presence or absence of OVA.

The Cooperative Interaction of Th and B Cells is MHC Restricted

Using Th cells derived from radiation induced chimeras, it has been proposed that Th to B cell interactions are not themselves MHC restricted, but rather, that Th to antigen presenting cell interactions are MHC restricted, while Th to B cell interactions are not (13). We have taken advantage of the T cell clones shown in Table 1 to examine this question in the plaque forming cell response to the hapten phosphorylcholine *in vitro*, as well as the B cell proliferative response to OVA in the presence of the cloned T cells.

Table 1. The Proliferative Response of Cloned, OVA Specific T Cells is MHC Restricted

Stimulator Spleen Cells		In Vitro Proliferative Responses of Cloned T Cells			
		B6d ($H-2^d$)		D4b ($H-2^b$)	
Strain	H-2	No OVA	With OVA	No OVA	With OVA
BALB/c	d	70	<u>950</u>	220	270
BALB.B	b	90	95	190	<u>1100</u>

$2x10^3$ cloned, OVA specific T cells from BALB/c (B6d) or BALB.B (D4b) were stimulated with $2x10^5$ mitomycin C treated spleen cells with or without 200 μg/ml in 200 μl cultures for 5 days. They were pulsed for 3 hours with ^{3}H--thymidine and harvested. Results are expressed as mean cpm incorporated of triplicate cultures; standard errors are less than 20% and have been omitted.

These studies are presented in detail elsewhere (14); they are summarized in Table 2. The experiment involved mixing B cells from two MHC disparate strains, BALB/c ($H-2^d$) and C57BL/6 ($H-2^b$), and examining the plaque forming cell response to PC-OVA generated by one or the other of the helper T cell clones shown in Table 1. In both cultures, plaque forming cell responses were obtained. The source of the antibody forming cells in each case was determined at the time of the plaque assay by killing the plaque forming cells with anti-K^b monoclonal antibody and complement. Since all cultures contained plaque forming cells, it is clear that both $H-2^b$ and $H-2^d$ accessory cell function, if necessary for such responses, was present in all cultures. However, the plaque forming cells in each case were MHC identical to the Th clone used in that culture. Furthermore, a mixture of the two clones gave rise to responses in both B cell populations (data not shown). If restriction in such responses acted at the Th to macrophage level, then cultures to which cells from a single clone are added should have had plaque forming cells from both MHC types in roughly equal numbers. If, on the other hand, restriction operated at the Th to B cell level, the result obtained here would be expected. We conclude from such experiments that Th cells are MHC restricted in their interaction with B cells directly. This, in turn, provides a distinct role for Ia antigens on the surfaces of B lymphocytes, namely, to direct the interaction of helper T cells with B cells. Since helper T cells are I region restricted, as are these cloned Th cells (see below), it seems likely that antigen-specific B cells bind PC-OVA, which is recognized on the B cell surface in association with I region gene products. The I region gene products involved in this interaction have been shown to be the Ia antigens (15). Since Ia

Table 2

OVA-Specific Th Clone	Anti-PC Antibody Response to PC-OVA	
	PFC/10^6 B Cells	% PFC Bearing $H-2^b$
B6d ($H-2^d$)	153	< 10%
D4b ($H-2^b$)	191	89%

10^5 cloned, OVA-specific T cells were added to a total of $3x10^6$ B cells prepared from normal spleen by treatment with anti-Thy-1, anti-Lyt-1, and anti-Lyt-2 plus rabbit complement. No response was obtained in the absence of Th. Antigen was PC-OVA at 0.1 μg/ml. Plaques were assayed on day 5 of response, and the PFC bearing $H-2^b$ determined by killing PFC with complement alone or with monoclonal anti-K^b antibody plus complement. % PFC corrected for kills on syngeneic mixtures of Th and B cells. Details of the analysis of cloned helper T cells are described previously (10,11).

antigens are expressed mainly on B lymphocytes, Th cells will interact selectively with B lymphocytes as a direct consequence of the I region restriction of Th cells.

Alloreactive T Cells Can Be Antigen Specific, MHC Restricted Helper T Cells

We have previously demonstrated that isolated populations of alloreactive T cells can be specifically immunized with protein antigens in nude mouse adoptive hosts. Such cells interestingly lack helper T cells acting selectively on T15 idiotype bearing B cells in the response to PC-OVA (16). Such helper T cell populations are also deficient in the second helper T cell observed early in the adoptive secondary antibody response to DNP-OVA, as seen by a shift in the slope of the dose response curve of log antibody concentration v. log helper T cell number, given in Table 3. The slope of this line is directly related to the number of interacting helper T cells present in the population. Thus, in both the anti-PC and the anti-DNP antibody response in vivo, helper T cells specific for OVA could be demonstrated in populations of T cells highly enriched for reactivity to a particular alloantigen ($H\text{-}2^b$), and in each instance evidence was obtained that a second population previously characterized as being dependent on Ig for its development was not found in this population.

Table 3. Isolated, Alloreactive T Cells Immunized with OVA in B Mice Lack a Helper T Cell Required for the Early Anti-DNP Antibody Response to DNP-OVA

Source of Helper T Cells: B Mice Given	Slope of the Helper T Cell Titration Curve	
	Day 6	Day 12
Normal BALB/c T Cells	2.1	1.13
BALB/c Anti-$H\text{-}2^b$ T Blasts	1.1	1.26

Helper T cells primed with OVA from adult-thymectomized, lethally irradiated, bone marrow reconstituted mice given lg-gradient purified BALB/c anti-C57BL/6J T cell blasts (16) or normal BALB/c T cells and immunized one month later with OVA were transformed to irradiated, syngeneic recipients along with $2x10^7$ DNP-KLH primed B cells (anti-Thy-1.2 and C treated spleen cells) and boosted with DNP-OVA, 5μg/mouse. Mice were bled 6 and 12 days later, anti-DNP antibody determined by Farr assay, and the results plotted on log-log paper. The slope of the Th dose response curve was determined from these linear plots (2).

This result has now been confirmed for the response to PC-OVA by using cloned helper T cell lines (10,11). Specifically, the helper T cell lines A3d and D4b can be shown to be alloreactive in that they respond to spleen cells from mice bearing particular non-self MHC encoded antigenic determinants in the absence of OVA. These same T cell clones respond to OVA in the presence of syngeneic spleen cells. Finally, these T cell clones, when cultured with syngeneic B cells and PC-OVA, induce the production of anti-PC antibody, and this antibody response is not dominated by the T15 idiotype produced by about 90% of plaque forming cells found during in vivo anti-PC antibody responses (10,11). These results are shown in Tables 4 and 5.

Table 4. OVA-Specific, MHC Restricted Cloned T Cell Lines Respond Specifically to Non-Self MHC Encoded Antigens

Stimulator Spleen Cells		Proliferative Response of Cloned T Cell Lines			
Strain	H-2	A3d ($H-2^d$)		D4b ($H-2^b$)	
		No OVA	+ OVA	No OVA	+ OVA
BALB.B	b	2700	3900	320	<u>4600</u>
BALB/c	d	5500	<u>40000</u>	400	450
BALB.K	k	3600	4800	<u>2900</u>	<u>3200</u>
BIO.G	q	<u>26000</u>	<u>26000</u>	220	180

Experiment carried out as in Table 1.

Table 5. OVA-Specific, MHC Restricted, Alloreactive Cloned T Cell Lines Help PC-Specific B Cells Respond to PC-OVA with Low T15 Idiotype Expression

Anti-PC PFC Response	OVA-Specific T Cell Clone Added to B Cells		
	B6d	A3d	D4b
Plaques per 10^6 B Cells	75	173	275
Percent Idiotype Positive	17%	12%	24%

Experiment carried out as in Table 2, except only syngeneic B cells were added to the cloned Th cells. Percent idiotype positive PFC determined by adding monoclonal anti-T15 idiotype antibody (10,11).

T Cells Raised in the Absence of Ig or B Cells Lack a Helper T Cell that is Required for Optimal Production of Anti-DNP and Anti-PC Antibody

As mentioned previously, helper T cells derived from mice treated from birth with anti-mouse μ chain antibody such that the mice were agammaglobulinemic lacked a helper T cell required for high early secondary anti-DNP antibody responses as well as for the dominant production of the T15 idiotype in anti-PC responses. This finding led to the proposal that there were two distinct classes of helper T cells, one of which is MHC recognizing and the other of which is Ig recognizing (7,17). The selective activation of idiotype-bearing B cells by such T cells would seem to argue strongly for this point. The finding that cloned helper T cells, which are antigen-specific and MHC restricted, lead to an anti-PC plaque forming cell response that is low in idiotype is also consistent with this division of cells. One of us (KB) has also demonstrated that the helper T cell that acts selectively on idiotype bearing B cells is not MHC restricted (18,19). In both the anti-DNP and the anti-PC systems, evidence has been obtained strongly supporting the concept that such helper T cells are antigen (that is, carrier) specific but do not require a hapten-carrier bridge in order to act on a B cell (5,8,9). A schematic diagram of these findings is given in Figure 1.

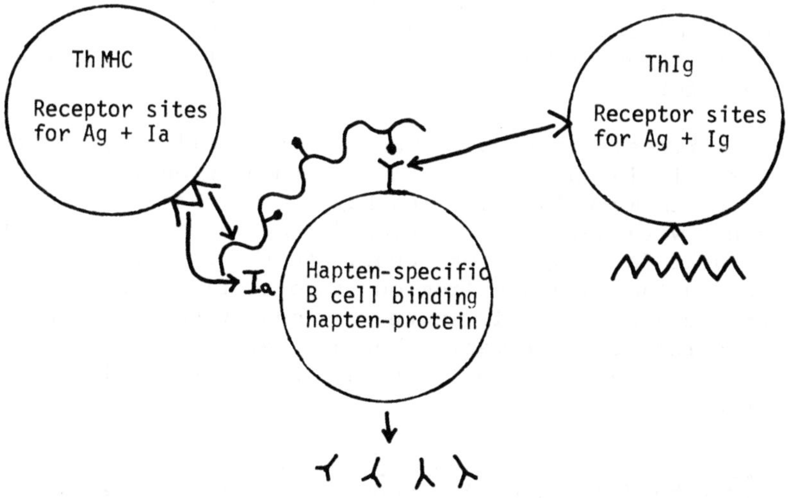

Figure 1. Two types of protein antigen specific helper T cells interact with an antigen-binding B cell. One recognizes self Ia on the B cell surface, the other a self Ig determinant on the B cell surface.

Thus, our studies have shown that optimal responses to hapten-carrier conjugates, and optimal production of dominant idiotypes, require the synergistic action of two protein-specific helper T cells. One of these appears to recognize self immunoglobulin determinants on B cells, but only acts on such cells if the helper T cell has itself been stimulated by carrier. Since the carrier need not bear the hapten, it seems likely that the carrier in this case is not necessarily presented by the B cell. The B cell also can only be stimulated to produce antibody if it has been activated by a second helper T cell that recognizes carrier, bound to the B cell's antigen-specific immunoglobulin receptor molecule, in the context of the B cell's Ia antigens.

The immunoglobulin recognizing T helper cell requires immunoglobulin and/or B cells to be present during its development (7-9). In the absence of self Ig molecules, such cells can not be detected. It should be mentioned that not all idiotype-specific helper T cells will necessarily have these properties. There is evidence from several laboratories, including our own, for a second type of anti-self idiotypic helper T cell. In some studies (20-24), anti-self idiotypic T cells are raised by direct immunization with idiotype in adjuvant; these cells can act directly on B cells to regulate the production of antibody (22-24). Recently, evidence that such cells may play a role in other B cell idiotype responses has been obtained by Cerny and colleagues (25,26). Such cells appear to be distinct from the protein-specific, self immunoglobulin recognizing T cells described here in that they are controlled by MHC-linked Ir genes (23) and are active in T-independent B cell responses (25,26). They can also appear to activate B cells in the total absence of exogenous antigen (22,26). The dependence of such cells on self immunoglobulin during development is now being tested in our laboratories.

MHC-restricted, Ia-recognizing helper T cells, by contrast, develop normally in the absence of self immunoglobulin (27). However, studies in many laboratories have shown that such cells require contact with self Ia molecules during their development in order that self Ia recognition occur. In the following sections, we will review the evidence from our laboratories that Ia antigens are the products of specific immune response genes, and we will also discuss recent experiments showing that the amount of an Ir gene product present on a cell can have pronounced effects on the funtioning of that product.

<u>Antigen-Specific T Cell Responses Which Are Controlled by Ir Genes Are Inhibited by Monoclonal Antibody to Ia Antigens Mapping to the Same I Subregion as the Ir Gene</u>

We have prepared a number of monoclonal anti-Ia antibodies, and we have used these antibodies to probe the function of Ia antigens. Of particular interest to the present discussion are the antibodies Y-17 and Y-8P. Y-17 is a monoclonal anti-A_eE_α antibody that will react with $A_e{}^{b,k,r,s}$ chains associated with $E_\alpha{}^{d,k,p,r}$ chains. It reacts with neither chain in isolation. These results have been obtained by serological, FACS and biochemical analysis, and have been published previously (28). Y-8P is a monoclonal anti-I-$A^{b,f,p,q,r,s}$ antibody, which also reacts with I-A^k at very low dilutions. The difference in titer in killing H-2^b and H-2^k targets, as well as 2-dimensional gel electrophoresis, have allowed us to map the genes encoding the determinant recognized by Y-8P to the I-A subregion. Both monoclonal antibodies react with B cells, but do not react with T cells.

We have previously published the finding that Y-17 could specifically inhibit the response of B10.A(5R) mice, which express $A_e{}^bE_\alpha{}^k$ complexes and react with Y-17, to the antigen poly-(L-glutamic acid, L-lysine, L-phenylalanine) (GLPhe), an antigen the response to which is controlled by two genes mapping in the I-A and I-E subregions respectively. Y-17 also inhibited the response of B10.A ($A_e{}^kE_\alpha{}^k$) and B10.S(9R) ($A_e{}^sE_\alpha{}^k$) mice to the antigen pigeon cytochrome c, the response to which is also controlled by genes mapping in the I-A and I-E subregions. These findings have been confirmed by blocking the response of clones specific for pigeon cytochrome c and for GLPhe with Y-17 (29,30). Y-17 and Y-8P have also been shown to block mixed lymphocyte reactions directed at A_eE_α and $A_\alpha A_\beta$ respectively, in the appropriate strain combinations (31). Y-17 will not inhibit the T cell proliferative response to the antigen TGAL, the response to which is controlled by the I-A subregion; however, killing the antigen presenting cells with Y-17 and complement decreases responses by T cells to all antigens to the same degree, independent of the subregion controlling the response (28). Thus, antigen-presenting cells bear both $A_\alpha A_\beta$ and A_eE_α complexes, but these complexes function independently in presenting antigen to T cells. This is also true in blocking mixed lymphocyte reactions, in which the responders and stimulators differ at both $A_\alpha A_\beta$ and A_eE_α; Y-17 and Y-8P both inhibit partially, and when added together to the cultures, their effects are additive (31). Thus, Ia antigens, whether serving as self or as non-self specificities, are expressed independently on the surfaces of stimulating cells.

We have also shown that monoclonal antibody Y-8P can specifically block T cell clone responses to antigen in the context of self I-A antigens as well as cloned T cell responses to non-self I-A encoded antigens. These results are presented in Table 6, in which H-2^d clones are not blocked by Y-8P because it does not react with I-A^d, but the responses of these clones to the non-self I-A antigen I-A^p is blocked, while the H-2^b clone D4b is blocked in its response to OVA in the context of self I-A^b by Y-8P.

Table 6. Inhibition of the Proliferative Response of Cloned, OVA-Specific, Alloreactive T Cell Lines to OVA plus Self I-A or to Non-Self I-A Encoded Antigens

Stimulator Spleen Cells		Y-8P	Response of T Cell Clone to Antigen			
Strain	H-2		A3d (H-2^d)		D4b (H-2^b)	
			No OVA	+ OVA	No OVA	+ OVA
BALB/c	d	-	600	3500	300	300
BALB/c	d	+	-	2900	-	-
C57BL/6	b	-	400	400	300	4800
C57BL/6	b	+	-	-	-	480
B10.F(13R)	p*	-	4900	4200	340	370
B10.F(13R)	p*	+	1500	-	-	-

Experiment was performed as in Table 1. Y-8P added at 1:500 final concentration. *B10.F(13R) is identical in the MHC to H-2P except at H-2^p, where it is D^b.

These data taken together have led us to the following conclusions about Ir genes, Ia antigens, antigens stimulating mixed lymphocyte reactions, and restriction elements in T-macrophage and T-B collaboration:

1. Ia glycoproteins serve as lymphocyte activating deteriminants in mixed lymphocyte reactions, as restriction elements in T-macrophage and T-B interactions, and are the functional products of immune response genes. This conclusion is based on the ability of monoclonal anti-Ia antibodies to quantitatively inhibit these responses, and the identical mapping of the genes encoding the Ia antigens with the Ir genes. It is supported both by studies of I region mutant mice (32,22) and by studies on the quantitation of Ia antigens and Ir gene effects (see below; 34, 35).

2. As far as we can determine, all Ir genes and all restriction elements thus far described can be attributed to products of the I-A and I-E subregions. As described by Baxevanis et al. (36), at least the defining antigens of the I-B subregion can be attributed to an interaction between I-A and I-E gene products. Similar tests on other Ir effects mapping to the I-B subregion may eventually eliminate this subregion from I region maps. Furthermore, Sprent et al. (37), using monoclonal anti-I-A plus Y-17 *in vivo* could completely inhibit the priming of helper T cells specific for the complex antigen sheep red blood cells at normal immunizing doses. This strongly suggests that all the restriction elements recognized by helper T cells in association with this complex nominal antigen can be attributed to I-A and I-E gene products.

3. These results provide strong evidence for the concept that Ia antigens are the products of Ir genes, but they do not suggest a mechanism by which Ir genes function through these molecular products. Why, for instance, does $A_e^bE_\alpha^k$ present GLPhe but not pigeon cytochrome c, while $A_e^sE_\alpha^k$ presents both, and $A_e^kE_\alpha^k$ presents pigeon cytochrome c but not GLPhe?

If the four polypeptide products of I-A (A_α, A_β, and A_e) and I-E (E_α) account for all of the functions of immune response genes, we are left with the problem of accounting for the apparent specificity of Ir gene effects. In the next section we will describe evidence that further supports the conclusion that Ir gene products are Ia antigens, and that they function by specifically associating with antigen on the surface of the antigen presenting cell. Those Ir gene products that produce sufficiently stable complexes with the nominal antigen would then be of the responder type, while those that do not form such complexes would be of the non-responder type. This has been proposed by us (38) and by others previously (39,40). It has still to be proven at the level of the antigen-presenting cell surface.

The Amount of Ia Antigen Expressed on Antigen Presenting Cells Can Determine Responder or Non-Responder Status

Nominal antigens, such as pigeon cytochrome c, the response to which is under the control of A_eE_α complexes, make ideal probes for immune response gene function. The E_α chain is relatively invariant biochemically, but only certain strains of mice synthesize such chains. In the absence of an E_α chain, no A_eE_α complexes can be found on the cell surface. The A_e chain is highly polymorphic. Thus, it has been found that responsiveness to pigeon cytochrome c requires the functioning of two genes, one coding for an appropriate A_e chain ($A_e^{k,s}$) and one encoding E_α. If one prepares cloned T cell lines specific for pigeon cytochrome c from B10.A mice ($A_e^kE_\alpha^k$), spleen cells from any strain expressing A_e^k complexed with E_α chains of any haplotype will present antigen to these clones, while B10.D2, which has $A_e^dE_\alpha^d$ complexes will not. However, spleen cells from mice having A_e^k in the presence of A_e^u and E_α^u ((B10.A(4R) x B10.PL)F1) will not present pigeon cytochrome c to such clones at conventional optimal stimulating doses of antigen. In exploring this surprising finding, it was demonstrated that F1 hybrid mice between H-2^u parental strains and strains having $A_e^{b,k,s}$ but no E_α chain would preferentially express the $A_e^uE_\alpha^u$ complex on the cell surface, and would express relatively little of the $A_e^{b,k,s}E_\alpha^u$ complexes. This quantitative defect in cell surface Ia antigen expression appears to be due to the preferential association of A_e^u chains with E_α^u chains, thus precluding cell surface expression of A_eE_α complexes of other types in such cells. Table 7 summarizes data from several different types of analysis relating the amount of Ia antigen on the surface

Table 7. Relationship of the Amount of Ia Antigen Detected by Monoclonal Antibody Y-17 to Functional Status as a Responder to Pigeon Cytochrome C

Strain Tested	Haplotype of Origin A_eE_α [1]	Ratio % Cytotoxicity Y-17:Anti-Ia.7[2]	Relative Amount $A_e^kE_\alpha$		Predicted % Cytotoxicity[5]	Response to Pigeon Cytochrome c
			Absorption[3]	FACS[4]		
B10.A	k:k	90%	100	100	85%	High
(B10.A(4R)xB10.D2)	k:d;d:d	60%	60	40	50%	Intermediate
(B10.A(4R)xB10.PL)	k:u;u:u	0%	12	10	(0%)	Non-responder
B10.A(4R)	k:-	-	0	0	-	Non-responder
B10.PL	u:u	0%	0	0	0	Non-responder

1. The first letter designates the origin of A_e, the second the origin of E_α chains; underlined complexes are known to react with Y-17 from biochemical experiments.
2. Percent kill with Y-17 divided by percent kill with anti-Ia.7 antibody.
3. The amount of $A_e^kE_\alpha$ complexes on the cell surface determined by quantitative absorption analysis relative to the homozygous strain B10.A.
4. The amount of $A_e^kE_\alpha$ complexes on the cell surface determined by quantitative analyses on the fluorescence-activated cell sorter (FACS) relative to B10.A.
5. The percent kill is predicted from curves obtained with the FACS, and from the finding that (B10.A(4R)xB10.PL)F1 spleen cells are not significantly killed by Y-17; thus, cells with fluorescence intensities in the range of those seen with (B10.A(4R)xB10.PL)F1 will not be killed by Y-17, while those with greater intensity will be killed.

of spleen cells with their Ir gene function. All the quantitative studies of the amount of Ia antigen expressed were carried out using the monoclonal antibody Y-17. It can be seen by direct cytotoxicity, by absorption analysis, by FACS analysis, and by Ir gene function, that the amount of Ia antigen expressed per cell is directly related to responder or non-responder status. Such data also explain the so-called "cis-trans" effects in complementing Ir gene function (41), since F1 mice have significantly less Ir gene product expressed than homozygotes.

If Ir gene products are Ia antigens, and if they function by association with the nominal antigen (here pigeon cytochrome c) on antigen presenting cell surfaces, then one might expect that increasing the dose of antigen in (B10.A(4R)xB10.PL) F1 mice would lead to an immune response. This was tested *in vitro* using cloned B10.A T cells and antigen presenting cells of various strains. Since the T cells were cloned, alloreactivity could be avoided. In this system, it was found that increasing the dose of antigen does indeed lead to effective antigen presentation by (B10.A(4R)xB10.PL) F1 spleen cells. The relative amount of Ia antigen as detected by

Table 8. Dose of Pigeon Cytochrome c Required for Half Maximal Stimulation of Cloned B10.A T Cells Specific for Pigeon Cytochrome c with Different Sources of Antigen Presenting Cells

Antigen Presenting Cells			
Strain	Haplotype of Origin of A_e:E_α	Relative Amount of Y-17*	Half-Optimal Dose (Relative)[o]
B10.A	k:k[+]	100	1
B10.A(4R)	k:-	0	>100
(B10.A(4R)xB10.D2)F1	k:d;d:d	50	1.5
B10.D2	d:d	0	>100
(B10.A(4R)xB10.PL)F1	k:u;u:u	10	15
B10.PL	u:u	0	>100

*See Table 7 data derived from quantitative absorption analyses and from FACS analysis and averaged.

\+ A_eE_α complexes capable of reacting with Y-17 are underlined.

o Half optimal dose determined by titration of pigeon cytochrome c into cultures containing cloned T cells from B10.A mice and irradiated spleen cells from the various strains, and measuring proliferation on day 3. All dose response curves in which a response was obtained led to the same optimal response, but at different doses of antigen. A result of >100 means no response was observed at the highest dose of antigen tested.

Y-17 and the amount of pigeon cytochrome c required for half optimal responses by these cloned T cells are given in Table 8. It can be seen that these two quantitites are inversely related. Furthermore, in preliminary experiments, raising the dose of pigeon cytochrome c used for in vivo immunization also leads to priming in the (B10.A(4R)xB10.PL) F1, but not in the B10.PL or B10.A(4R) homozygotes. Thus, these data suggest strongly that the Ir gene product, as detected by Y-17, which can completely inhibit the response of the B10.A cloned T cell line to pigeon cytochrome c with any source of stimulator, associates with the nominal antigen on the surface of the antigen presenting cell, and that one needs to generate sufficient complexes of Ir gene product and nominal antigen to trigger a T cell response. This reaction can be driven by increasing either the nominal antigen or the Ir gene product. Thus, these data, while not proving that such associations occur, are most strongly consistent with the hypothesis that they do, and with the notion that non-responsiveness is, in some cases, related to the failure to form such complexes.

CONCLUSIONS

The experiments summarized here have dealt with T cells of the helper type. The greatest precision obtainable in such experiments is obviously to use cloned helper T cells and purified B cells. Unfortunately, this is not possible in all cases. Furthermore, while it gives us a great deal of information regarding the interaction of T and B cells, as well as about the specificity of antigen recognition by T cells, it can not explain those characteristics of the interaction of foreign antigen with an intact immune system that determine the ultimate outcome of the encounter. Nevertheless, such experiments do teach us about the elements that are present, and about what they can do functionally. Within these constraints, the experiments discussed here lead one to certain general conclusions.

First, it is clear that the helper T cell that is MHC restricted and which recognizes antigen in the context of self Ia antignes is critical in antibody responses to protein antigens. This Th cell probably is activated initially by antigen in the context of an antigen presenting cell. However, once primed, it is perfectly able to interact in an Ia-restricted fashion with a B cell binding the antigen by means of its Ig receptor. Indeed, Gray and colleagues have recently used rabbit Ig as an antigen, and show that B cells can present this antigen to proliferating T cells when it is in the form of rabbit anti-MIg, that is, when it is bound to the B cell's Ig receptor (42). Clearly, this makes good sense. Helper T cells, to perform their function of activating antigen-specific B cells, must activate just those B cells that bind antigen in this way. Whether binding of antigen to the Ig on the B cell surface also alters the ability of the B cell to activate T cells is not known. Other cells binding antigen (for instance suppressor T cells), will not be acted on by these helper T cells because they lack the appropriate Ia antigens on their surface. The cell that activates suppressor T cells is not MHC restricted, but rather is restricted by Igh-linked gene products (43).

The activity of these helper T cells is modulated by the products of Ir genes. In the present studies, evidence has been presented that the Ia antigens are the products of these genes, and that these gene products can affect the intensity of an immune response either as a result of the presence or absence of a particular polymorphic form of the product, or as a result of the amount of a given form of the product present on the surfaces of antigen presenting cells. This latter finding is interpreted to imply a physical association between Ir gene product and nominal antigen, but such complexes have yet to be physically demonstrated in Ir-protein antigen systems. This implication that non-responsiveness reflects the failure to form such complexes is certainly strengthened by these findings, which are difficult to explain by other proposed mechanisms of Ir gene function not involving the formation of complexes.

Finally, we have briefly described the finding of a second class of protein antigen specific helper T cell that seems to interact with the B cell not via Ia antigens but rather by means of recognizing the B cell's Ig receptor. This T cell is not MHC restricted nor under the control of known Ir gene products. Its role seems to be in the rapid early activation of particular clones of B cells. It may also control the "quality" of an immune response by acting selectively on B cells bearing Ig of different isotypes; such cells have been described by Seman and colleagues (44). These T cells require Ig for their development in a way similar to the requirement for Ia antigens in the development of MHC restricted T cells.

Thus, B cell responses to antigens are regulated by at least two types of helper T cells. As discussed above, there will surely be further classes of helper T cells to be discovered and discriminated from those already well characterized. Furthermore, the activity of these cells is governed by an array of suppressor and even contrasuppressor T cells characterized by Gershon and his colleagues (45). Our present approach is to use the cloned helper T cells as targets for these suppressor and contrasuppressor T cells in order to determine the precise workings of a model immune system _in vitro_. Perhaps from such an approach, an understanding of the complexities of an _in vivo_ immune response and its regulation will emerge.

ACKNOWLEDGEMENTS

These studies were supported in part by the Howard Hughes Medical Institute, and by NIH grants AI-13766, AI-14579, AI-15132, AI-14349, and AI-17576.

REFERENCES

1. N. A. Mitchison. The carrier effect in the secondary response to hapten-protein conjugates. II. Cellular cooperation, Eur. J. Immunol. 1:18 (1971).
2. C. A. Janeway, Jr. Cellular cooperation during in vivo antibody responses. I. The effect of cell number, J. Immunol. 114:1394 (1975).
3. D. H. Katz, M. Graves, M. E. Dorf, H. DiMuzio and B. Benacerraf. Cell interactions between histoincompatible T and B lymphocytes. VII. Cooperative responses between lymphocytes are controlled by genes in the I region of the H-2 complex, J. Exp. Med. 141:263 (1975).
4. M. J. Bevan and P. J. Fink. The influence of thymus H-2 antigens on the specificity of maturing killer and helper cells, Immunol. Rev. 42:3 (1979).
5. D. H. Katz, T. Hamaoka, M.E. Dorf, P. H. Maurer and B. Benacerraf. Cell interactions between histoincompatible T and B lymphocytes. IV. Involvement of the immune response (Ir) gene in the control of lymphocyte interaction in responses controlled by the genes, J. Exp. Med. 138:734 (1973).
6. C. A. Janeway, Jr., D. L. Bert and F. W. Shen. Cell cooperation during in vivo anti-hapten antibody responses. V. Two synergistic $Lyl^{+}23^{-}$ helper T cells with distinctive specificities, Eur. J. Immunol. 10:231 (1980).
7. C. A. Janeway, Jr., R. A. Murgita, F. I. Weinbaum, R. Asofsky and H. Wigzell. Evidence for an immunoglobulin-dependent antigen-specific helper T cell, Proc. Nat. Acad. Sci. (USA) 74:4582 (1977).
8. K. Bottomly and D. E. Mosier. Mice whose B cells can not produce the T15 idiotype also lack an antigen-specific helper T cell required for T15 expression, J. Exp. Med. 150:1399 (1979).
9. K. Bottomly, C. A. Janeway, Jr., B. J. Mathieson and D. E. Mosier. Absence of an antigen-specific helper T cell required for the expression of the T15 idiotype in mice treated with anti-μ antibody, Eur. J. Immunol. 10:159 (1980).
10. K. Bottomly and F. Jones III. Idiotypic dominance manifested during a T-dependent anti-phosphorylcholine response requires a distinct helper T cell, in: "B Lymphocytes in the Immune Response," N. Klinman, D. Mosier, I. Scher and E. Vitetta, eds, Elsevier/North Holland, New York pg. 415 (1981).

11. K. Bottomly. Activation of the idiotypic network: environmental and regulatory influences, in: "Immunoglobulin Idiotypes," C. A. Janeway, Jr., E. Sercarz, H. Wigzell and C. F. Fox, ed., Academic Press, New York. In press.
12. B. Sredni, H. Y. Tse, C. Chin and R. H. Schwartz. Antigen-specific clones of proliferating T lymphocytes. I. Methodology, specificity and MHC restriction, J. Immunol. 126:341 (1981).
13. A. Singer, K. S. Hathcock and R. J. Hodes. Cellular and genetic control of antibody responses. V. Helper T-cell recognition of H-2 determinants on accessory cells but not B cells, J. Exp. Med. 149:1208 (1979).
14. B. Jones and C. A. Janeway, Jr. Cooperative interaction of B lymphocytes with antigen-specific helper T lymphocytes is MHC restricted, Nature (Lond) 292:547 (1981).
15. J. Sprent. Effects of blocking helper T cell induction in vivo with anti-Ia antibodies. Possible role of I-A/E hybrid molecules as restriction elements, J. Exp. Med. 152:996 (1980).
16. K. Bottomly and C. A. Janeway, Jr. Selected populations of alloreactive T cells contain helper T cells but lack ThId, an antigen-specific helper T cell required for dominant production of the T15 idiotype, Eur. J. Immunol. 11: (1981).
17. K. Bottomly and D. E. Mosier. Analogous dual specificity of helper T cells cooperating in the generation of clonally restricted antibody responses, in: "Strategies of Immune Regulation, E. Sercarz and A. Cunningham, ed., Academic Press, New York, pg. 487 (1980).
18. K. Bottomly and P. H. Maurer. Antigen-specific helper T cells required for dominant production of an idiotype (ThId) are not under immune response (Ir) gene control, J. Exp. Med. 152:1571 (1980).
19. K. Bottomly and D. E. Mosier. Antigen-specific helper T cells required for dominant idiotype expression are not H-2 restricted, J. Exp. Med. 154:411 (1981).
20. C. A. Janeway, Jr., N. Sakato and H. N. Eisen. Recognition of immunoglobulin idiotypes by thymus-derived lymphocytes, Proc. Nat. Acad. Sci. (USA) 74:2357 (1975).
21. N. Sakato, C. A. Janeway, Jr. and H. N. Eisen. Immune response of BALB/c mice to the idiotype of T15 and other myeloma proteins of BALB/c origin. Implications for an immune network and antibody multispecificity, Cold Spring Harbor Symp. Quant. Biol. 41:719 (1976).
22. K. Eichmann, I. Falk and K. Rajewsky. Recognition of idiotypes in lymphocyte interactions. II. Antigen-independent cooperation between T and B lymphocytes that possess similar and complimentary idiotypes, Eur. J. Immunol. 8:853 (1978).

23. T. Jorgensen, B. Bogen and K. Hannestad. Recognition of variable (V) domains of myeloma protein 315 by B- and T-lymphocytes, in: "Immunoglobulin Idiotypes", C. Janeway, E. Sercarz, H. Wigzell and C. F. Fox, eds., Academic Press, New York. In press.
24. R. G. Lynch, G. L. Milburn, R. G. Hoover, J. W. Rohrer and B. Dieckgraefe. Myeloma proteins are targets and inducers of immunoregulatory signals, in: "Immunoglobulin Idiotypes", C. Janeway, E. Sercarz, H. Wigzell and C. F. Fox, eds., Academic Press, New York. In press.
25. G. Kelsoe, D. Isaak and J. Cerny. Thymic requirement for cyclical idiotypic and reciprocal anti-idiotypic immune responses to a T-independent antigen, J. Exp. Med. 151:289, (1980).
26. J. Cerny and M. J. Caulfield. Stimulation of specific antibody-forming cells in antigen-primed nude mice by the adoptive transfer of syngeneic anti-idiotypic T cells, J. Immunol. 126:2262 (1981).
27. C. A. Janeway, Jr., B. Broughton, E. Dzierzak, B. Jones, D. D. Eardley, S. Durum, K. Yamauchi, D. R. Green and R. K. Gershon. Studies of T lymphocyte function in B cell deprived mice, in: "Immunoglobulin idiotypes and their expression," C. Janeway, E. Sercarz, H. Wigzell and C. F. Fox, eds., Academic Press, New York. In press.
28. E. A. Lerner, L. A. Matis, C. A. Janeway, Jr., P. P. Jones, R. H. Schwartz and D. B. Murphy, Monoclonal antibody against an Ir gene product, J. Exp. Med. 152:1085 (1980).
29. L. A. Matis, D. Lebwohl, M. Ultee, E. Margoliash and R. H. Schwartz. Gene complementation in the T lymphocyte response to pigeon cytochrome C, Fed. Proc. 39:1127 (1980).
30. B. Sredni, L. A. Matis, E. A. Lerner, W. E. Paul and R. H. Schwartz. Antigen-specific T cell clones restricted to unique F_1 major histocompatibility complex determinants. Inhibition of proliferation with a monoclonal anti-Ia antibody, J. Exp. Med., 153:677 (1981).
31. C. A. Janeway, Jr., E. A. Lerner, P. J. Conrad and B. Jones, The precision of self and non-self major histocompatibility complex encoded antigen recognition by cloned T cells, Behring Institute Mitteilungen. In press.
32. M. Michaelides, M. Sandrin, G. Morgan, IFC McKenzie, R. Ashman and R. W. Melvold. Ir gene function in an I-A subregion mutant B6.C-H-2^{bm12}, J. Exp. Med. 153:464 (1981).
33. L. J. Rosenwasser and B. T. Huber. The xid gene controls Ia.W39-associated immune response gene function, J. Exp. Med. 153:1113 (1981).
34. L. A. Matis, P. P. Jones, D. B. Murphy, S. M. Hedrick, E. A. Lerner, C.A. Janeway, Jr., J. M. McNicholas and R. H. Schwartz. Immune response gene function correlates with the cell surface expression of an Ia antigen. II. A

quantitative deficiency in A_eE_α complex expression causes a corresponding defect in antigen presenting cell function, J. Exp. Med., submitted
35. J. M. McNicholas, D. B. Murphy, L. A. Matis, R. H. Schwartz, E. A. Lerner, C. A. Janeway, Jr. and P. P. Jones. Immune response gene function correlates with the expression of an Ia antigen. I. Preferential association of certain A_e and E_α chains results in a quantitative deficiency in expression of an A_eE_α complex, J. Exp. Med., submitted.
36. C. N. Baxevanis, Z. A. Nagy and J. Klein. A novel type of T-T cell interaction removes the requirement for I-B region in the H-2 complex, Proc. Nat. Acad. Sci. (USA) 78:3809 (1981).
37. J. Sprent, E. A. Lerner, J. Bruce and F. W. Symington. Inhibition of T cell activation in vivo with mixtures of monoclonal antibodies specific for I-A and I-A/E molecules, J. Exp. Med. 154:188 (1981).
38. C. A. Janeway, Jr., H. Wigzell and H. Binz. Hypothesis: Two different VH gene products make up the T cell receptors, Scand. J. Immunol. 5:993 (1976).
39. B. Benacerraf. An hypothesis to relate the specificity of T lymphocytes and the activity of I region-specific Ir genes in macrophages and B lymphocytes, J. Immunol. 120:1809 (1978).
40. A. S. Rosenthal. Determinant selection and macrophage function in genetic control of the immune response. Immunol. Rev. 40:136 (1978).
41. R. H. Schwartz, C. F. Merryman and P. H. Maurer. Gene complementation in the T lymphocyte proliferative response to poly ($glu^{57}lys^{38}tyr^{5}$): evidence for effects of polymer handling and gene dosage, J. Immunol. 123:272 (1979).
42. R. W. Chestnut and H. M. Grey. Studies on the capacity of B cells to serve as antigen presenting cells, J. Immunol. 126:1075 (1981).
43. D. D. Eardley, F. W. Shen, H. Cantor and R. K. Gershon. Genetic control of immunoregulatory circuits. Genes linked to the Ig locus govern communication between regulatory T cell sets, J. Exp. Med. 150:44 (1979).
44. M. Seman and J. Morisset. Analysis of T clones provide evidence for two distinct populations of helper T cells (Th1 and Th2) one of them participating to the isotypic regulation of the antibody response, J. Supramolec. Struct. and Cell Biochem. Suppl. 5:86 (1981).
45. R. K. Gershon, D. D. Eardley, S. Durum, D. R. Green, F.-W. Shen, K. Yamauchi, H. Cantor and D. B. Murphy. Contrasuppression. A novel immunoregulatory activity, J. Exp. Med. 153:1533 (1981).

5

DISSECTION OF THE MOLECULAR PARAMETERS FOR T-CELL RECOGNITION OF A MYOGLOBIN ANTIGENIC SITE

Colin R. Young and M. Zouhair Atassi

Department of Immunology, Mayo Medical School

Rochester, MN 55905

SUMMARY

Previous studies from this laboratory have resulted in the determination of the antigenic structure of sperm-whale myoglobin (Mb). In the present work, we have investigated the fine specificity requirements for T-cell recognition of one of the Mb antigenic sites (antigenic site 5). The antigenic site (peptide 145-153) and seven progressively longer peptides, increasing in length stepwise by two residues at a time up to 22 residues in length (peptide 132-153), were synthesized. In addition, four truncated peptides were synthesized with intentional deletions at Tyr-151 and Ala-144. The T-cell recognition of these purified synthetic peptides was examined here in detail in three strains of mice (BALB/cByJ, B10.D2/n and SJL/J). Mb-primed mice afforded T-cells which proliferated to smaller peptides (two or four residues longer than the site; i.e. peptides 145-153 and 143-153) and more so to the longer peptides 135-153 and 132-153 and to Mb. No response was obtained to the truncated peptides, thus underscoring the fine specificity of T-cells. No response was obtained also to intermediate-sized peptides. The latter result, due to an unfavorable mode of folding, suggested a conformational dependency in T-lymphocyte recognition and was confirmed by additional studies employing peptide priming. Thus priming with either the 11- or 22-residue peptides afforded T-cells which proliferated to both of these two peptides, but not to intermediate-sized peptides.

+Abbreviations: Mb, myoglobin; PBS, 0.01 M sodium phosphate buffer, pH 7.2, containing 0.15 M NaCl; PPD, purified protein derivative.

More importantly, peptide-primed cells did not proliferate on challenge in vitro with Mb. This indicated that the in vivo recognition (priming) was to the unfolded peptide and these cells will not therefore recognize native Mb in vitro. Tolerization of neonatal mice with native Mb, which was total for the native protein at the T-cell level, had no effect on in vivo recognition (priming) and in vitro proliferation due to the unfolded peptide. Conversely, tolerization by any unfolded peptide had little or no effect on in vivo recognition of, and proliferation to, native Mb. It was concluded that T-cell recognition of native proteins (or at least of Mb) is dependent on protein conformation. A peptide may not cause proliferation of native protein-primed T-cells even though it contains an antigenic site. Finally, macrophage presentation of protein antigens is not related to processing via fragmentation of antigen by macrophage.

INTRODUCTION

Elucidation of the molecular and cellular parameters that control and regulate the immune response to multi-valent complex protein antigens has been a major effort of this laboratory since our determination of the antigenic structure of sperm-whale myoglobin (Mb)+ (1). The antibody response to this protein was found to be under the control of H-2 linked Ir-genes (2-4). Studies on the T-lymphocyte proliferative response to Mb in congenic and recombinant strains indicated that the T-cell response is under H-2 linked Ir-gene control similar to the humoral response (5). With the knowledge of the entire antigenic structure of Mb, it was possible now to examine the genetic control of the antibody and T-lymphocyte response to the antigenic sites of Mb. The native protein has five antigenic sites (1) located in exposed conformationally sensitive continuous portions of the polypeptide chain. The primary structure and conformational locations of these antigenic sites are shown in Figures 1 and 2. Recent studies have also shown that recognition of the antigenic sites is not dependent on the immunized species. The same five antigenic sites on Mb are recognized by antisera raised in rabbit, goat, pig, cat, chicken and outbred (6) and inbred mouse (4). The sites recognized by mouse B cells are also recognized by T-cells (4,7).

The genetic control of the T-lymphocyte response to the five synthetic antigenic sites of Mb was determined in independent and recombinant strains of mice (7). Studies with the recombinant strains suggested that Ir-genes to sites 1 and 2 map in the I-A and I-C subregion (see Figure 1, Chapter 6, this volume). Ir-genes to sites 4 and 5 mapped only in the I-A subregion, whereas response to site 3 is controlled by a non-H-2 linked gene. Previously it was thought that site 1 and 2 may be controlled by a single Ir-gene, but we have recently shown that sites 1 and 2 are indeed controlled by separate Ir-genes (8).

Site 1	15 16 21 22 (Ala)-Lys-Val-Glu-Ala-Asp-Val-(Ala)	6 (or 7)
Site 2	56 62 Lys-Ala-Ser-Glu-Asp-Leu-Lys	7
Site 3	94 99 Ala-Thr-Lys-His-Lys-Ile	6
Site 4	113 119 His-Val-Leu-His-Ser-Arg-His	7
Site 5	145 146 151 (Lys)-Tyr-Lys-Glu-Leu-Gly-Tyr	6 (or 7)

Figure 1. Primary structures of the five antigenic sites of sperm-whale Mb. Residues in parentheses are part of the antigenic site only with some antisera. Thus, for Site 1, the reactive region invariably occupies sequence 16-21 and with some antisera alanine 15 is part of the region (which will then correspond to sequence 15-21) while with other antisera alanine 22 is an essential part of the region (which will then correspond to sequence 16-22). This site occupies either 6 or 7 residues depending on antiserum. For Sites 2, 3 and 4 no such 'displacement' or 'shift' to one side or the other has been observed (at least with the antisera so far studied). In the case of Site 5, lysine 145 can be part of the reactive region only with some antisera and this site will therefore comprise 6 or 7 residues, depending on the antiserum (see reference 1).

These T-cell proliferation studies to Mb and its antigenic sites were carried out using a peritoneal exudate T-lymphocyte enriched cell (PETLES) assays (9). However, there are certain drawbacks in using the PETLES assays: 1) The assay is time consuming since the cells are assayed for proliferation three weeks after the priming immunization; 2) Because of the very poor yield of T-cells obtained after nylon wool column fractionation of the peritoneal exudate cells, a large number of mice is required for each proliferation experiment; 3) Because of this poor yield of cells, only a small number of synthetic peptides can be tested at any one time by in vitro proliferation assays; 4) Low stimulation indices and Δc.p.m. are obtained using the PETLES assay, especially with small peptides.

Because of these inherent problems in the PETLES assay, we decided to investigate the T-lymphocyte proliferative response to Mb, its synthetic antigenic sites and synthetic peptides in high responder H-2^d and H-2^s strains of mice using a lymph node (Tail) assay (10,11). We were specifically interested in determining whether there was an optimum size of synthetic peptide that

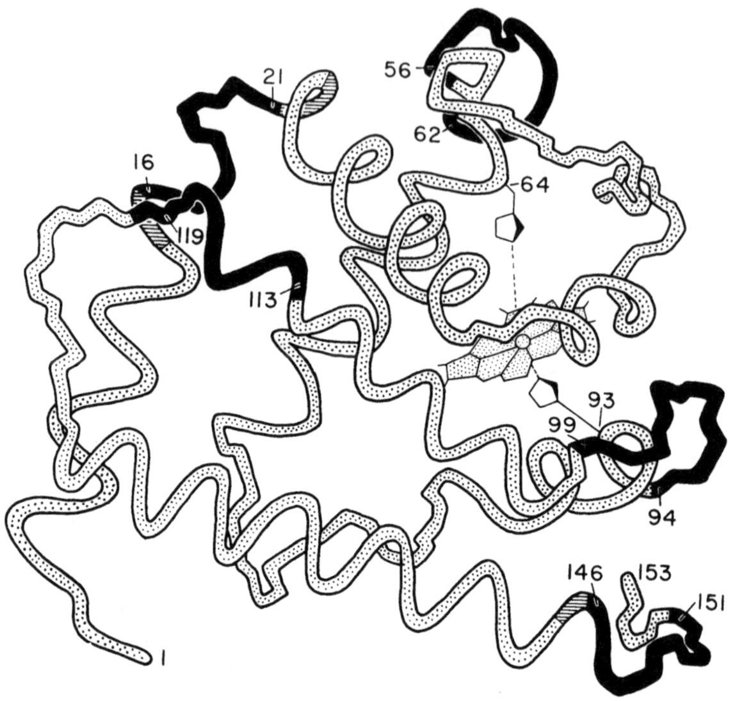

Figure 2. A schematic diagram showing the mode of folding of Mb and its antigenic structure. The solid black portions represent segments which have been shown to comprise accurately entire antigenic reactive regions. The striped parts, each corresponding to one amino acid residue only, can be part of the antigenic reactive region with some antisera. The dotted portions represent parts of the molecule which have been shown exhaustively to reside outside reactive regions (see reference 1).

stimulated T-cell proliferation in the Tail assay. For example, site 5 (peptide 145-151) of only 7 amino acids causes a weak T-cell proliferative response by Tail assay. If we now increase the size of this synthetic peptide, is it possible to obtain a stronger proliferative response? We investigated this using synthetic peptides carrying antigenic site 5 but increasing in length in steps of two amino acid residues at a time to the left of the site (Figure 3). Site 5 was chosen for this study because of its structural position in the native Mb molecule (Figure 2).

Site 5	145 (Lys)-Tyr-Lys-Glu-Leu-Gly-Tyr 151
Peptide 145-153	145 Lys-Tyr-Lys-Glu-Leu-Gly-Tyr-Gln-Gly 153
Peptide 143-153	143 Ala-Ala-Lys-Tyr-Lys-Glu-Leu-Gly-Tyr-Gln-Gly 153
Peptide 141-153	141 Asp-Ile-Ala-Ala-Lys-Tyr-Lys-Glu-Leu-Gly-Tyr-Gln-Gly 153
Peptide 139-153	139 Arg-Lys-Asp-Ile-Ala-Ala-Lys-Tyr-Lys-Glu-Leu-Gly=Tyr-Gln-Gly 153
Peptide 137-153	137 Leu-Phe-Arg-Lys-Asp-Ile-Ala-Ala-Lys-Tyr-Lys-Glu-Leu-Gly-Tyr-Gln-Gly 153
Peptide 135-153	135 Leu-Glu-Leu-Phe-Arg-Lys-Asp-Ile-Ala-Ala-Lys-Tyr-Lys-Glu-Leu-Gly-Tyr-Gln-Gly 153
Peptide 132-153	132 Asn-Lys-Ala-Leu-Glu-Leu-Phe-Arg-Lys-Asp-Ile-Ala-Ala-Lys-Tyr-Lys-Glu-Leu-Gly-Tyr-Gln-Gly 153

Figure 3. Primary structure of the synthetic peptides of Mb. The residue in parenthesis is not required as part of the antigenic site in all antisera (1). The synthesis and purification of site 5 has been described previously (13).

Since site 5 occurs at the carboxy terminal end of the molecule, it is possible to increase the size of the synthetic peptides carrying site 5 in one direction only, a procedure which is not possible for any of the other four antigenic sites of Mb (see Figure 2). For this study peptides were synthesized which corresponded to the following regions in the native Mb molecule; peptide 145-151 (antigenic site 5), peptide 145-153, peptide 143-153, peptide 141-153, peptide 139-153, peptide 137-153, peptide 135-153 and peptide 132-153 (Figure 3). It should be noted that peptide 132-153, although made synthetically, corresponds to an antigenically active cyanogen bromide cleavage fragment of Mb (12).

In addition to investigating the T-lymphocyte proliferative response to these synthetic peptides carrying antigenic site 5, we also examined the fine specificity of T-cell recognition of peptides. It is known that antibodies to native Mb are very discriminating in their recognition of synthetic peptides. For example, deletion of tyrosine at position 151 abrogates the antibody recognition of peptide 145-151 (antigenic site 5) (13,1). Furthermore, since mouse T-cells recognize the sites on Mb that are recognized by B-cells (4,7), we were interested in determining whether T-cells are as discriminating as B-cells in their recognition of Mb antigenic sites. In order to determine this fine specificity of recognition by T-cells, we synthesized synthetic peptides of Mb which contained deliberate amino acid deletions. The amino acids deliberately deleted in these peptides are Tyr-151 and Ala-144. The peptides used in this study were truncated peptides 145-153, 143-153, 141-153 and 139-153 (Figure 4).

Another point of interest concerning the cellular recognition of Mb and its synthetic peptides is whether T-cell recognition is dependent in any way on the conformation of the antigenic region that is recognized. It has long been assumed that T-cell receptors recognize sequence features or unfolded or denatured portions of protein antigens and that B-cell receptors recognize conformational features. The studies on T-cell recognition of the aforementioned peptides together with those on neonatal mice that were tolerized with either Mb or synthetic peptides and the T-cell recognition of these tolerized mice have revealed that native conformation plays a role in the T-cell recognition of Mb.

MATERIALS AND METHODS

Animals

Eight to twelve-week-old male mice of strains BALB/cByJ, B10.D2/n and SJL/J were purchased from Jackson Laboratories, Bar Harbor, Maine. Breeding pairs of BALB/cByJ strain were also purchased from Jackson Laboratories.

Truncated Peptide 145-153	Lys-Tyr-Lys-Glu-Leu-Gly—Gln-Gly (145 … 153)	
Truncated Peptide 143-153	Ala—Lys-Tyr-Lys-Glu-Leu-Gly—Gln-Gly (143 … 153)	
Truncated Peptide 141-153	Asp-Ile-Ala—Lys-Tyr-Lys-Glu-Leu-Gly—Gln-Gly (141 … 153)	
Truncated Peptide 139-153	Arg-Lys-Asp-Ile-Ala—Lys-Tyr-Lys-Glu-Leu-Gly—Gln-Gly (139 … 153)	

Figure 4. Primary structure of the truncated synthetic peptides of Mb. The continuous lines shown at positions 144 and 151 indicate the absence of Ala and Tyr residues respectively at these positions

Preparation of Myoblogin

Sperm-whale Mb used in these studies was the major chromatographic component (No. 10) obtained by CM-cellulose chromatography of twice crystallized Mb (14). The major chromatographic component was homogeneous by polyacrylamide disc gel electrophoresis.

Synthesis of Peptides

The peptides were synthesized by the solid-phase method. Synthesis of the peptides, cleavage from the resin, deprotection and purification by ion-exchange chromatography on CM-Sephadex were carried out by the procedures previously described in detail (13,15). The purity of each peptide was confirmed by peptide mapping. Each gave a single spot in heavily loaded maps, indicating purities of 98% or better, and its amino acid composition was in excellent agreement with that expected from its covalent structure.

Two groups of peptides were synthesized. The first group corresponds to the Mb sequences: 145-151 (antigenic site 5), 145-153, 143-153, 141-153, 139-153, 137-153, 135-153 and 132-153. The primary structures of these synthetic peptides are shown in Figure 3.

The second group of peptides, designated truncated peptides, were synthesized with specific amino acid deletions. The amino acids deleted in these peptides were tyrosine at position 151 and alanine at position 144. These truncated peptides correspond to the Mb sequence 145-153, 143-153, 141-153 and 139-153. The primary structures of these synthetic peptides are shown in Figure 4.

Immunization with Mb

Mice were immunized with 50 μg of Mb in 100 μl of a 1:1 (v/v) emulsion of antigen in phosphate-buffered saline (PBS) and complete Freund's adjuvant (Difco) containing mycobacterium strain H37Ra, subcutaneously into the tail (10,11). Seven days after priming, the draining lymph nodes were collected and proliferation assays carried out.

Immunization with Synthetic Peptides

Mice were immunized subcutaneously into the tail with 100 μg of synthetic peptide in 100 μl of a 1:1 (v/v) emulsion of peptide in PBS and complete Freund's adjuvant containing mycobacterium strain H37Ra. Seven days after priming, the draining lymph nodes were collected and proliferation assays carried out.

Tolerance to Mb or Synthetic Peptides

24-Hour old mice were tolerized by injection with either 1000 μg of Mb or synthetic peptide in 100 μl of PBS in a single intraperitoneal site. Eight weeks later, mice were immunized into the tail with either 50 μg of Mb or synthetic peptide in 100 μl of 1:1 (v/v) emulsion of antigen in PBS and complete Freund's adjuvant containing mycobacterium strain H37Ra. Seven days after priming, the draining lymph nodes were collected and proliferation assays carried out.

Lymph Node Cell Proliferation Assays

Seven days post priming, the draining inguinal and paraaortic lymph nodes were collected. Single cell suspensions were prepared in RPMI 1640 supplemented with 10% fetal calf serum (v/v), 100 I.U./ml each of penicillin and streptomycin, 2mM glutamine, 25mM HEPES (all reagents were obtained from Gibco) and 3×10^{-5} M 2-mercaptoethanol (Sigma). Cells (3×10^5 per well) were co-cultured with PPD (positive control antigen), a synthetic peptide (16) that does not represent any part of Mb (nonsense peptide, see Table 5) and synthetic peptides (test antigens) in flat bottom microtiter wells (Falcon No. 3042) in a reaction volume of 200 μl. After three days of culture, [^{3}H]-thymidine (Research Products International) (1 μCi in 10 μl) was added per well. Twenty-four hours later, cells were harvested onto filter paper discs. The discs were dried and the amount of radioactivity determined. Results are expressed as Δcpm and are calculated:

Δcpm = (cpm of cultures containing test or positive control antigen) -

(cpm of cultures containing nonsense peptide)

cpm of cultures containing medium alone or nonsense peptide were virtually identical.

RESULTS

Immunization with Mb

It is shown (Table 1) that following immunization with Mb in high responder H-2^d and H-2^s strains of mice the T-cells proliferate in vitro to Mb and peptides 145-153, 143-153, 135-153 and 132-153. No T-cell proliferation was observed with the intermediate size peptides, namely peptides 141-153, 139-153 and 137-153. It is of interest that in vitro T-cell proliferation is obtained with the smaller and large synthetic peptides, but not with the intermediate size peptides, even though the latter peptides each contain the same antigenic site (1). It is noteworthy that none

Table 1

Strain distribution of lymphocyte proliferation responses to in vitro challenge with Mb and synthetic peptides, following immunization with Mb.

[a]Challenge antigen or peptide	[b]Δcpm for BALB/cByJ cultures	[b]Δcpm for B10.D2/n cultures	[b]Δcpm for SJL/J cultures
Peptide 145-153	1452	1914	3237
Peptide 143-153	2604	2848	6190
Peptide 141-153	0	0	0
Peptide 139-153	0	0	0
Peptide 137-153	0	0	0
Peptide 135-153	18195	3777	15337
Peptide 132-153	3174	5835	6664
Mb			
Truncated Peptide 145-153	0	0	0
Truncated Peptide 143-153	0	0	0
Truncated Peptide 141-153	0	0	0
Truncated peptide 139-153	0	0	0

[a]Final concentration of antigen or peptide used was 100 μg/ml.

[b]The standard deviations of the cpm did not exceed ± 15%.

of the truncated peptides 145-153, 143-153, 141-153 or 139-153 stimulated in vitro T-cell proliferation.

Immunization with Synthetic Peptides

It is shown (Table 2) that following immunization with peptide 145-151 (antigenic site 5) in H-2^d and H-2^s mice the T-cells do not proliferate in vitro to Mb, site 5 or any of the synthetic peptides 145-153, 143-153, 141-153, 139-153, 137-153, 135-153 or 132-153. The T-cells were viable and capable of in vitro proliferation as indicated by the high Δcpm obtained with PPD. This may have been due to a critical size requirement for effective priming by peptide antigens, or due to the conformational dependency of T-cell recognition (i.e. unfolding that occur in peptides), or due to minimum size requirement for binding to T-cell receptors. Further studies were designed to investigate these questions.

Since no in vitro T-cell proliferation was obtained following immunization with a single synthetic antigenic site of only 7 amino acids, we decided to investigate the in vitro T-cell proliferation after immunization with larger synthetic peptides.

Table 2

Strain distribution of lymphocyte proliferation responses to *in vitro* challenge with Mb and synthetic peptides, following immunization with peptide 145-151 (antigenic site 5).

[a]Challenge antigen or peptide	[b]Δcpm for BALB/cByJ cultures	[b]Δcpm for B10.D2/n cultures	[b]Δcpm for SJL/J cultures
Peptide 145-151 (site 5)	0	0	0
Peptide 145-153	0	0	0
Peptide 143-153	0	0	0
Peptide 141-153	0	0	0
Peptide 139-153	0	0	0
Peptide 137-153	0	0	0
Peptide 135-153	0	0	0
Peptide 132-153	0	0	0
Mb	0	0	0
PPD	88,973	84,228	70,946

[a]Final concentration of antigen or peptide used was 100 μg/ml.

[b]The standard deviations of the cpm did not exceed ± 15%.

Following immunization with peptide 143-153 in $H\text{-}2^d$ and $H\text{-}2^s$ mice it was found (Table 3) that T-cells proliferate to peptides 145-153, 143-153, 137-153, 135-153 and 132-153. No *in vitro* T-cell proliferation was observed to Mb or the intermediate size peptides 141-153 or 139-153. Of interest is the finding that *in vitro* proliferation is obtained with the smaller and larger peptides, but not with the intermediate size peptides. No *in vitro* T-cell proliferation was obtained with any of the truncated peptides 145-153, 143-153, 141-153 or 139-153.

Following immunization with peptide 132-153 in high responder $H\text{-}2^d$ and $H\text{-}2^s$ mice it was found (Table 4) that T-cells proliferate *in vitro* to peptides 145-153, 137-153, 135-153 and 132-153. No T-cell proliferation was obtained to Mb or the intermediate size peptides 141-153 and 139-153, as was observed in immunization with peptide 143-153. Also, no *in vitro* proliferation was obtained with any of the truncated peptides 145-153, 143-153, 141-153, 139-153.

Table 3

Strain distribution of lymphocyte proliferation responses to *in vitro* challenge with Mb and synthetic peptides, following priming with peptide 143-153.

[a]Challenge antigen or peptide	[b]Δcpm for BALB/cByJ cultures	[b]Δcpm for B10.D2/n cultures	[b]Δcpm for SJL/J cultures
Peptide 145-153	2581	2960	2646
Peptide 143-153	2621	0	1980
Peptide 141-153	0	0	0
Peptide 139-153	0	0	0
Peptide 137-153	0	3183	4013
Peptide 135-153	6032	3861	6631
Peptide 132-153	8516	6610	6100
Mb	0	0	0
PPD	67557	60214	87988
Truncated Peptide 145-153	0	0	0
Truncated Peptide 143-153	0	0	0
Truncated peptide 141-153	0	0	0
Truncated peptide 139-153	0	0	0

[a]Final concentration of antigen or peptide used was 100 μg/ml.

[b]The standard deviations of the cpm did not exceed ± 15%.

Tolerance to Mb or Synthetic Peptides

One of the central points in investigating tolerance to Mb and its synthetic peptides is the ability to demonstrate that it is possible to tolerize with native Mb. We have found that it is indeed possible to tolerize mouse T-cells to Mb since these cells do not proliferate *in vitro* following a single intraperitoneal injection of Mb in PBS in neonatal mice and subsequent challenge with Mb in complete Freund's adjuvant (Table 5). It is shown that mice are tolerized to Mb since no *in vitro* T-cell proliferation is observed to Mb or any of its synthetic peptides. The T-cells were viable and capable of *in vitro* proliferation as indicated by the high Δcpm obtained with PPD.

Another important point to determine is whether this tolerance in the Mb system is specific or non-specific. This was investigated by tolerizing mice with the truncated peptide 139-153. It is shown in Table 6 that neonatal mice injected with truncated peptide 139-153 and subsequently immunized with Mb in complete Freund's adjuvant are indeed tolerant to this truncated peptide.

Table 4

Strain distribution of lymphocyte proliferation responses to in vitro challenge with Mb and synthetic peptides, following immunization with peptide 132-153.

[a]Challenge antigen or peptide	[b]Δcpm for BALB/cByJ cultures	[b]Δcpm for B10.D2/n cultures	[b]Δcpm for SJL/J cultures
Peptide 145-153	3001	2283	2892
Peptide 143-153	0	2169	0
Peptide 141-153	0	0	0
Peptide 139-153	0	0	0
Peptide 137-153	3065	2578	7541
Peptide 135-153	10086	8053	20055
Peptide 132-153	32241	30625	31727
Mb	0	0	0
PPD	53745	60239	44546
Truncated Peptide 145-153	0	0	0
Truncated Peptide 143-153	0	0	0
Truncated Peptide 141-153	0	0	0
Truncated Peptide 139-153	0	0	0

[a]Final concentration of antigen or peptide used was 100 µg/ml.

[b]The standard deviations of the cpm did not exceed ± 15%.

Table 5

Lymphocyte proliferation responses to in vitro challenge with Mb and synthetic peptides. Neonatal BALB/cByJ were tolerized to Mb in PBS, and eight weeks later immunized with Mb in complete Freund's adjuvant.

[a]Challenge antigen or peptide	[b]Δcpm
Peptide 143-153	0
Peptide 135-153	0
Peptide 132-153	0
Mb	0
PPD	42499
[c]Nonsense peptide	0
Truncated peptide 139-153	0

[a]Final concentration of antigen or peptide used was 100 µg/ml.

[b]The standard deviations of the cpm did not exceed ± 15%.

[c]Nonsense peptide is: Glu-Ser-Ser-Gly-Thr-Gly-Ile. This peptide was previously designed and synthesized (16) so that it does not resemble any part of the Mb molecule.

Furthermore this tolerance is specific to the truncated peptide 139-153 since the T-cells proliferate in vitro to Mb and peptides 143-153, 135-153 and 132-153. Thus it appears that tolerizing mice to a truncated peptide does not affect the T-cell recognition of Mb or its antigenic sites.

In Table 7 it is shown that following tolerance with Mb and subsequent immunization with peptide 132-153 in complete Freund's adjuvant, the T-cells are tolerant to Mb. However, these T-cells proliferate to peptides 143-153, 135-153 and 132-153. Thus it appears that T-cells can proliferate to these synthetic peptides of Mb even though they have been tolerized to the corresponding regions of these peptides in the native Mb. No in vitro T-cell proliferation was obtained with the truncated peptide 139-153.

Mice tolerized to peptide 143-153 and subsequently immunized with Mb in complete Freund's adjuvant are tolerant to the peptide 143-153, but are not tolerant to Mb (Table 8). This tolerance, although induced by the injection of neonatal mice with peptide 143-153, is also effective against peptides 135-153 and 132-153. Thus tolerance to a small synthetic peptide of only eleven amino acids, is effective against larger peptides containing the same antigenic site.

In Table 9 it is shown that following tolerance with peptide 132-153 and subsequent immunization with Mb in complete Freund's adjuvant, the T-cells are tolerant to peptide 132-153, but are not tolerant to Mb. Also, these tolerant T-cells are non-responsive to peptides 135-153 and 143-153. Thus tolerance to a large synthetic peptide of 21 amino acids is effective against smaller peptides containing the same antigenic site. Interestingly it should be noted that these T-cells, although made tolerant to peptide 132-153 show a significant reduction in their proliferative response to Mb (for comparison see Table 6). Thus tolerance with a large peptide (of only 22 amino acids) containing only 1 antigenic site is effective in reducing the proliferative response to the native Mb (of 153 amino acids) by some 47%.

DISCUSSION AND CONCLUSIONS

T-Cell Response of Mb-Primed Mice

The first aspect of these studies is concerned with T-cell recognition in Mb-primed animals. T-cells from these animals responded slightly to peptide 145-153 and the response improved on lengthening the peptide by two residues (i.e. peptide 143-153). However, the total loss in proliferative response on further stepwise lengthening which was observed in peptides 141-153, 139-153 and 137-153 was unexpected. The reappearance of a strong response on elongation of the peptide by only two more residues (peptide

Table 6

Lymphocyte proliferation responses to in vitro challenge with Mb and synthetic peptides. Neonatal BALB/cByJ mice were tolerized with truncated peptide 139-153 in PBS, and eight weeks later immunized with Mb in complete Freund's adjuvant.

[a]Challenge antigen or peptide	[b]Δcpm
Peptide 143-153	1197
Peptide 135-153	4635
Peptide 132-153	7090
Mb	17308
PPD	36767
[c]Nonsense peptide	0
Truncated peptide 139-153	0

[a]Final concentration of antigen or peptide used was 100 μg/ml.

[b]The standard deviations of the cpm did not exceed ± 15%.

[c]See footnote in Table 5.

Table 7

Lymphocyte proliferation responses to in vitro challenge with Mb and synthetic peptides. Neonatal BALB/cByJ mice were tolerized with Mb in PBS, and eight weeks later immunized with peptide 132-153 in complete Freund's adjuvant.

[a]Challenge antigen or peptide	[b]Δcpm
Peptide 143-153	1713
Peptide 135-153	10134
Peptide 132-153	25604
Mb	0
PPD	28132
[c]Nonsense peptide	0
Truncated peptide 139-153	0

[a]Final concentration of antigen or peptide used was 100 μg/ml

[b]The standard deviation of the cpm did not exceed ± 15%.

[c]See footnote in Table 5.

Table 8

Lymphocyte proliferation responses to in vitro challenge with Mb and synthetic peptides. Neonatal BALB/cByJ mice were tolerized to peptide 143-153 in PBS, and eight weeks later immunized with Mb in complete Freund's adjuvant.

[a]Challenge antigen or peptide	[b]Δcpm
Peptide 143-153	0
Peptide 135-153	0
Peptide 132-153	0
Mb	13398
PPD	38872
[c]Nonsense peptide	0
Truncated peptide 139-153	0

[a]Final concentration of antigen or peptide used was 100 μg/ml.

[b]The standard deviations of the cpm did not exceed ± 15%.

[c]See footnote in Table 5.

Table 9

Lymphocyte proliferation responses to in vitro challenge with Mb and synthetic peptides. Neonatal BALB/cByJ mice were tolerized to peptide 132-153 in PBS, and eight weeks later immunized with Mb in complete Freund's adjuvant.

[a]Challenge antigen or peptide	[b]Δcpm
Peptide 143-153	0
Peptide 135-153	0
Peptide 132-153	0
Mb	8823
PPD	34485
[c]Nonsense peptide	0
Truncated peptide 139-153	0

[a]Final concentration of antigen or peptide used was 100 μg/ml.

[b]The standard deviations of the cpm did not exceed ± 15%.

[c]See footnote in Table 5.

135-153) was also very surprising. Thus, it was evident that some intermediate-size peptides, even though they contain fully the antigenic region present in the shorter peptides, are unable to bind to T-cell receptors. This is analogous to the situation previously observed in binding of some Mb antigenic sites with antibody. In those studies, it was found (17,18) that lengthening the antigenic site by an additional neighboring sequence (1 to 5 residues) may result in appreciable steric obstruction in antibody binding due to an unfavorable mode of folding of some longer peptides in solution. This would suggest that binding of antigen with T-cell receptors is subject to similar steric and conformational considerations as the binding with antibody. The unfavorable mode of folding would disappear upon further elongation of the peptide. Long-range and short-range interactions will progressively play an increased role as the peptide gets longer and this improves its ability to approach the shape of the region it occupies in the intact protein (12, 19-21).

The above findings should serve to caution that the inactivity of a given peptide fragment by T-cell proliferation does not necessarily indicate the absence of an antigenic site within the peptide. For example, it has hitherto been customary when determining T-cell recognition sites of a protein molecule to cleave that molecule into fragments. These fragments are then tested for their proliferative capacity by *in vitro* cultures of T-cells from animals that had been primed with the whole molecule. If no proliferation is obtained to a specific fragment of the molecule, it is concluded that this fragment does not contain a T-cell recognition site. As we have shown here, such a conclusion could be incorrect since it is possible to obtain no proliferation with a peptide even though it does contain an antigenic site.

It is significant to note here that the lack of response by Mb-primed T-cells to any of the truncated peptides clearly underscores the fine discriminating ability of T-cells. Evidently the recognition requirements of T-cells are no less specific than those of B-cells since these deletions also obliterate ability of the peptide to bind antibody (13).

T-Cell Response of Peptide-Primed Mice

The conformational dependency of T-cell recognition and response suggested by the above experiments were further investigated by peptide-priming studies. Since priming with site 5 (peptide 145-151) failed to generate a proliferative response to Mb or any of the synthetic peptides, including site 5 itself, it was not clear whether this lack of response was due to a peptide-size requirement in T-cell priming. Therefore, mice were primed with peptide 143-153 (11 residues) or peptide 132-153 (22 residues). Priming with either of these two peptides afforded

T-cells that responded in each case very well to peptides 132-153, 135-153, 137-153 and gave a lower but significant response to the shorter peptides 145-153 and 143-153. The lack of response to the truncated peptides again underscores the fine specificity of the T-cells. Again, the lack of response to the intermediate-size peptides 141-153 and 139-153 suggested that these peptides possessed a mode of folding which differed from the other peptides and which was unfavorable for binding with T-cell receptors. The lack of response to Mb was intriguing and suggested a high level of conformational dependency for T-cell recognition.

It may appear at first glance contradictory that Mb-primed T-cells respond to peptides such as 143-153 and 132-153, whereas when these two peptides are used for priming the cells will not respond to Mb. But this is exactly what should be expected if conformational factors played a role in T-cell recognition. Thus, when Mb is the priming antigen, the region presently under study (antigenic site 5), would have a specific shape which is recognized by the T-cells. When these T-cells are then challenged _in vitro_ by the peptide, the receptors on the T-cells are able, by virtue of the conformational flexibility and freedom of the peptide, to induce the peptide into a conformation which is favorable for binding. The time-average conformation of a peptide may be predominantly unfolded, with only a small fraction of the conformational states being able at any given time to bind with T-cell receptors. However, the existing conformational equilibrium will generate more of that conformational state which is being taken out of the equilibrium by T-cell binding. In the reverse situation where the peptide is the priming antigen, the structure that will be recognized will be the time-average unfolded shape of the peptide. On challenge _in vitro_, the T-cells will then be expected to respond only to the unfolded structure. Because of the stability of the intact Mb molecule, the T-cells will not be able to induce it to present the region of site 5 in an unfolded manner similar to that used for educating the cells. The peptide-primed T-cells will not then be expected to respond to Mb. This is indeed what was observed.

These findings permit us to conclude that antigen processing by fragmentation does not precede, or is at least distinct from, macrophage presentation of antigen to T-cells. Protein antigen is recognized intact by T-cells. If the macrophage fragmented the Mb molecule and released active peptides containing antigenic sites, or factors such as genetically related factors (GRF's), proliferation would be expected to Mb following immunization with peptides 143-153 and 132-153.

Tolerization with Mb or Peptides

The role played by native conformation in Mb T-cell recognition was further confirmed by tolerance studies. Thus after tolerization with native Mb it will be expected that free peptides, because their unfolded shape is different from the conformation of that region within intact Mb, will continue to be recognized when used as priming and challenging antigens. On the other hand, tolerization with unfolded peptides should have little or no effect on recognition of intact Mb when the protein is used in tolerized mice as priming and challenging antigen. These were indeed exactly the observed findings. Therefore, the results of T-cell tolerance indicate that protein conformation plays an important role in this phenomenon as well in the normal T-cell recognition of native proteins.

ACKNOWLEDGEMENTS

We thank Nina Bren for superb technical assistance. This work was supported by grants AM-18920 and AI-18657 from the National Institutes of Health, U.S. Public Health Service.

REFERENCES

1. M.Z. Atassi, Antigenic structure of myoglobin: The complete immunochemical anatomy of a protein and conclusions relating to antigenic structures of proteins. Immunochemistry 12:423 (1975).
2. C.R. Young, A. Ebringer and J. Archer, Antigen dose and strain variation as factors in the genetic control of the immune response to sperm-whale myoglobin. Immunology 34:571 (1978).
3. J. Berzofsky, Genetic control of the immune response to mammalian myoglobins in mice. I. More than one I-region gene in H-2 controls the antibody response. J. Immunol. 120:360 (1978).
4. S.S. Twining, C.S. David and M.Z. Atassi, Genetic control of the immune response to myoglobin. IV. Mouse antibodies in outbred and congenic strains against sperm-whale myoglobin recognize the same antigenic sites that are recognized by antibodies raised in other species. Mol. Immunol. 18:447 (1981).
5. K. Okuda, P.R. Christadoss, S. Twining, M.Z. Atassi and C.S. David, Genetic control of the immune response to sperm-whale myoglobin in mice. I. T-lymphocyte proliferative response under H-2-linked Ir gene control. J. Immunol. 121:866 (1978).
6. S.S. Twining, H. Lehmann and M.Z. Atassi, The antibody response to myoglobin is independent of the immunized species: Analysis in terms of replacements in the antigenic sites and in environmental residues of the cross-reactions of fifteen myoglobins with sperm-whale myoglobin antisera raised in different species. Biochem. J. 191:681 (1980).

7. K. Okuda, S.S. Twining, C.S. David and M.Z. Atassi, Genetic control of the immune response to sperm-whale myoglobin in mice. II. T-lymphocyte proliferative response to the synthetic antigenic sites. J. Immunol. 123:182 (1979).
8. C.R. Young and M.Z. Atassi, Genetic control of the immune response to myoglobin. IX. Overcoming genetic control of antibody response to antigenic sites by increasing the dose of antigen used in immunization. J. Immunogen., in press (1982).
9. R.H. Schwartz, L. Jackson and W.E. Paul, T-lymphocyte enriched murine peritoneal exudate cells. I. A reliable assay for antigen-induced T-lymphocyte proliferation. J. Immunol. 115:1330 (1975).
10. G. Corradin, H.M. Etlinger and J.M. Chiller, Lymphocyte specificity to protein antigen. I. Characterization of the antigen induced in vitro T-cell dependent proliferative response with lymph node cells from primed mice. J. Immunol. 119:1048 (1977).
11. S.S. Alkan, Antigen-induced proliferation assay for mouse T-lymphocytes. Response to a monovalent antigen. Europ. J. Immunol. 8:112 (1978).
12. M.Z. Atassi and B.J. Saplin, Immunochemistry of sperm-whale myoglobin. I. The specific interaction of some tryptic peptides and of peptides containing all the reactive regions of the antigen. Biochemistry 7:688 (1968).
13. J. Koketsu and M.Z. Atassi, Immunochemistry of sperm-whale myoglobin. XVIII. Final delineation to ultimate boundaries of the single reactive region in sequence 120–153 by study of synthetic peptides. Biochim. Biophys. Acta 328:289 (1973).
14. M.Z. Atassi, Properties of components of myoglobin of the sperm whale. Nature (Lond.) 202:496 (1964).
15. J. Koketsu and M.Z. Atassi, Immunochemistry of sperm-whale myoglobin. XVI. Accurate delineation of the single region in sequence 1-55 by immunochemical studies of synthetic peptides: Some conclusions concerning antigenic structures of proteins. Immunochemistry 11:1 (1974).
16. S.S. Twining and M.Z. Atassi, Use of immunoadsorbents for the study of antibody binding to sperm-whale myoglobin and its synthetic antigenic sites. J. Immunol. Methods 30:139 (1979).
17. J. Koketsu and M.Z. Atassi, Immunochemistry of sperm-whale myoglobin. XIX. Accurate delineation of the single reactive region in sequence 56-87 by study of synthetic peptides. Biochim. Biophys. Acta 342:21 (1974).
18. M.Z. Atassi and R.-C. Pai, Immunochemistry of sperm-whale myoglobin. XXII. Accurate delineation of the single reactive region in sequence 103-120 by immunochemical studies of synthetic peptides. Immunochemistry 12:741 (1975).
19. R.M. Epand and H.A. Scheraga, The influence of long-range interactions on the structure of myoglobin. Biochemistry 7:2864 (1968).

20. R.P. Singhal and M.Z. Atassi, Conformational studies on modified proteins and peptides. II. Conformation of peptides with intact and overlapping helices obtained by cleavage of myoglobin at proline peptide bonds. Biochemistry 9:4252 (1970).
21. M.Z. Atassi and R.P. Singhal, Conformational studies on modified proteins and peptides. III. Conformation of peptides obtained by cleavage of myoglobin at arginine peptides bonds. J. Biol. Chem. 245:5122 (1970).

GENETIC CONTROL OF THE IMMUNE RESPONSE TO PROTEINS AND PEPTIDES

6

GENETIC CONTROL AND INTERSITE INFLUENCES ON THE IMMUNE RESPONSE TO SPERM WHALE MYOGLOBIN

Chella S. David and M. Zouhair Atassi

Dept. of Immunology, Mayo Medical School

Rochester, MN 55905

SUMMARY

Determination of the precise antigenic structure of sperm-whale myoglobin (Mb) has enabled us to focus our attention on the molecular and cellular factors that control and regulate the immune responses to the protein antigen. Our studies have shown that the immune responses to sperm-whale Mb are controlled by genes in the I region of the major histocompatibility complex (H-2) of mice. More importantly, the responses to the synthetic antigenic sites are each under separate genetic control. The recognition of the antigenic sites by antibodies is independent of the immunized species and of the time the antisera are obtained after the initial immunization (from nine days up to a year). The same sites are recognized by antisera raised in rabbit, goat, pig, cat, chicken and outbred and inbred mice. The same sites recognized by mouse B-cells are also recognized by mouse T-cells. No meaningful genetic control of antibody affinity was observed. Autoimmune antibody and T-lymphocyte proliferative responses were readily generated by immunizing an animal with self-Mb. With mouse Mb, the autoimmune T-lymphocyte response was under genetic control and mapped with the I-A and the H-2D end of the H-2 gene complex. In other recent studies we have shown, using several Mb variants, that the binding capacity of an antigenic site is fully accounted for by substitutions in the antigenic sites (actual contact residues) and in residues close (within 7.Å) to the sites (i.e. environmental residues). The

---

+Abbreviations: Mb, myoglobin; Ir-gene, immune response gene; H-2, the major histocompatibility complex; PBS, 0.01 M sodium phosphate buffer, pH 7.2, containing 0.15 M NaCl.

overall response to Mb is regulated by inter-site influences which can either be of a cooperative (help) nature or of a suppressive nature. Finally, genetic control of the responses to individual antigenic sites on a protein is not only determined by the genetic constitution of the host but also by the chemical properties of the individual sites. The H-2 subregions mapping the responses to given antigenic sites can also recognize other sites, which were previously unrecognizable in a homologous protein, if the chemical properties of these sites are suitably altered.

INTRODUCTION

Immune response studies using natural protein antigens have added a new dimension to the understanding of the mechanisms of immune recognition and responses (1). Studies using sperm whale myoglobin (Mb)$^{+}$ have extended the probes even further by defining discrete immune response genes to distinct antigenic sites on a multi-determinant protein molecule (2). This protein is composed of a single polypeptide chain comprising 153 amino acid residues (3) folded in a highly helical compact structure (4). The entire antigenic structure of sperm-whale Mb has been determined (2). The native protein has five continuous antigenic sites occupying the following conformationally distinct surface regions: Site 1 - sequence 15-22; site 2 - sequence 56-62; site 3 - sequence 94-99; site 4 - sequence 113-119; and site 5 - sequence 145-151. Determination of the antigenic structure of sperm whale Mb (2) has enabled us to focus our attention on the molecular and cellular parameters that control and regulate the immune responses to protein antigens. Extensive studies have been carried out using the synthetic antigenic sites as well as Mb variants from different species and this article will present only a brief outline of these studies.

GENETIC CONTROL OF THE IMMUNE RESPONSE TO MYOGLOBIN AND ITS ANTIGENIC SITES

Immune Response to Sperm Whale Myoglobin in Independent Haplotypes

Inbred strains of mice representing different H-2 haplotypes in different genetic backgrounds were tested for their ability to respond to the whole sperm-whale Mb (5). The Mb employed in these studies was the homogeneous major chromatographic component No. 10 (MbX) obtained by chromatography on CM cellulose (6). Eight mice of each strain were immunized with Mb (50 μg) in PBS (50 μl) mixed 1:1 (v/v) in complete Freund's adjuvant containing 1 mg/ml of microbacterium Tuberculosis in the two hind footpads and the animals were sacrificed three weeks later. The peritoneal exudate T-lymphocyte enriched cells (PETLES) and their in vitro cultures were prepared as described (7). Results from a representative strain distribution is shown in Table 1. H-2 haplotypes b and k were low responders regardless of the genetic backgrounds of the

Table 1

T-Lymphocyte Proliferative Response to Sperm-Whale Myoglobin in Independent H-2 Haplotypes[a]

Mouse Strains	H-2 Haplotype	cpm (Mean ± S.D.) Medium	Myoglobin (40 μg/ml	Stimulation Indices
B10	b	745 ± 82	2,034 ± 261	2.73
B6	b	276 ± 34	399 ± 106	1.45
B10.D2	d	1,046 ± 208	7,682 ± 1,452	7.34
DBA/2	d	1,362 ± 213	7,905 ± 438	5.80
BALB/cJ	d	815 ± 150	17,408 ± 2,481	21.35
B10.M	f	1,604 ± 187	8,142 ± 278	5.07
B10.WB	j	793 ± 156	1,056 ± 223	1.33
B10.BR	k	879 ± 122	730 ± 239	0.83
B10.K	k	730 ± 62	1,042 ± 94	1.43
C3H/HeJ	k	2,611 ± 407	2,449 ± 623	0.94
B10.P	p	1,208 ± 40	764 ± 117	0.63
B10.G	q	2,772 ± 196	3,489 ± 430	1.25
B10.RIII	r	2,125 ± 373	2,563 ± 369	1.20
B10.S	s	870 ± 539	6,278 ± 1,055	7.22
ASW	s	591 ± 65	6,529 ± 394	11.0

[a]About eight mice of each strain were injected with 50 μg of Mb with complete Freund's adjuvant in footpads. After 19 days, 2.5 x 10^5 PETLES were cultured with Mb at the final concentration of 40 μg/ml. After 3 days, [^{3}H]-thymidine-incorporated cells were harvested. PPD (40 μg/ml) response of each strain was between 15,432 and 128,016 cpm. Table is from Okuda et al. (8).

mice used. Haplotypes d, f and s were high responders. Within the d haplotype strains, BALB/c was a very high responder suggesting the influences of non-H-2 genes also in the response to sperm-whale Mb.

Mapping of Ir-Mb Genes

Results with the B10 congenic mice, B10.D2, B10, B10.M, B10.K and B10.S clearly showed that at least one gene controlling the response to Mb is linked to the major histocompatibility complex (5). In an attempt to map this gene, extensive studies with recombinant strains of mice were undertaken (5). In this review, we will merely show the critical strains which mapped the Ir-Mb genes (Table 2). Strains A.TL and A.TH differ only between the K and the D region, that is, in the I and S region. A.TL has the k alleles in the middle while A.TH has the s alleles. The aforementioned data had shown that k is a low responder and s is a high responder. A.TL was found to be a high responder while A.TH is a low responder, mapping the Ir-Mb genes between the K and D regions. B10.HTT which has the s alleles in K, A, B and J regions and the low responder k alleles in the rest of the regions was a high responder, whereas B10.S(8R) which has the low responder k alleles in the K and A regions and the high responder s alleles in the other regions was a low responder. The results of these two strains narrowed down the mapping of one Ir-Mb gene to the I-A subregion. This was confirmed for the d haplotype using D2.GD. Surprisingly, B10.A which has the high responder d alleles in the C, S and D regions and low responder k alleles elsewhere was also a high responder suggesting a second Ir-Mb gene involved in that control. Recent studies with cloned T cells have shown that this response in B10.A is due to the hybrid I-E molecule (A^kE^k) (Infante, Atassi and Fathman, this volume).

Immune Response to the Synthetic Antigenic Sites of Myoglobin

Having determined that the immune response to Mb is under genetic control, we proceeded to investigate (8) whether the responses to the individual sites are genetically controlled. In these experiments, we were asking the following questions: 1) Are all of the high responder mice responding to the same antigenic sites of sperm-whale Mb or to different sites? 2) Do the mice respond to more than one antigenic site? 3) Do any of the low responder mice respond to any single antigenic site , which was presented or recognized alone? To state it more concisely, are the responses to each antigenic site under separate genetic control? The animals were primed with whole Mb and then challenged in vitro with the different synthetic peptides representing the sites (Table 3). In addition, two types of control peptides were used: 1) Synthetic peptides of Mb representing non-antigenic regions; 2) a peptide which has no similarity to any part of the Mb molecule - nonsense peptide (Table 3). All independent haplotypes in every background were tested. We will discuss only the most representative results (Table 4). B10.D2 ($H\text{-}2^d$) responded to sites 1, 2, 3 and 5 whereas B10.S responded to sites 1,2 and 5. Hence, two high responders differ in the sites they respond to. Strain A.SW which is the same $H\text{-}2^s$ haplotype as B10.S also responded to site 3, suggesting that a non-H-2 linked gene controls the response to site 3. Among the non-responder haplotypes, B10 ($H\text{-}2^b$) and B10.K ($H\text{-}2^k$), a marginal response was seen for site 4. This suggests that in the absence of other sites,

Table 2

Genetic Mapping of the Ir Gene Controlling the T-Lymphocyte Proliferative Response to Sperm-Whale Myoglobin

	MHC Alleles									cpm (Mean ± S.D.)		
Mouse Strains	K	A	B	J	E	C	S	G	D	Medium	Myoglobin (40 μg/ml)	Stimulation Indices
A.TL	s	k	k	k	k	k	k	k	d	1,366 ± 130	1,614 ± 81	1.18
A.TH	s	s	s	s	s	s	s	s	d	1,907 ± 380	9,854 ± 526	5.16
B10.HTT	s	s	s	s	k	k	k	k	d	860 ± 118	9,801 ± 209	11.4
D2.GD	d	d	b	b	b	b	b	b	b	2,746 ± 135	11,670 ± 359	4.25
B10.A	k	k	k	k	k	d	d	d	d	1,272 ± 236	7,550 ± 467	5.94
B10.A(2R)	k	k	k	k	k	d	d	?	b	1,329 ± 603	5,831 ± 712	4.38
B10.A(5R)	b	b	b	k	k	d	d	d	d	675 ± 81	5,183 ± 467	7.89
B10.S(8R)	k	k	?	?	s	s	s	s	s	1,494 ± 286	2,570 ± 149	1.72
B10.S(9R)	s	s	?	k	k	d	d	d	d	463 ± 64	8,751 ± 2,140	18.9
C3H.OH	d	d	d	d	d	d	d	d	k	763 ± 103	4,556 ± 712	5.91
C3H.OL	d	d	d	d	d	d	k	k	k	529 ± 191	3,726 ± 1,431	7.00
ATFR.5	f	f	f	f	f	?	k	k	d	796 ± 146	9,352 ± 1,215	11.74
B10.A(4R)	k	k	b	b	b	b	b	b	b	761 ± 90	593 ± 133	0.78

[a]Methods are the same as in Table 1

Table is from Okuda et al. (8)

Table 3

Sequence of Synthetic Peptides

	Synthetic Antigenic Sites
Site 1	15 – 22 Ala-Lys-Val-Glu-Ala-Asp-Val-Ala
Site 2	56 – 62 Lys-Ala-Ser-Glu-Asp-Leu-Lys
Site 3	94 – 100 Ala-Thr-Lys-His-Lys-Ile-Pro
Site 4	113 – 120 His-Val-Leu-His-Ser-Arg-His-Pro
Site 5	145 – 151 Lys-Tyr-Lys-Glu-Leu-Gly-Tyr
Synthetic Control Non-Antigenic Regions	
Control 1	1 – 6 Val-Leu-Ser-Glu-Gly-Glu
Control 2	121 – 127 Gly-Asn-Phe-Gly-Ala-Asp-Ala
	Synthetic "Nonsense" Peptide
	Glu-Ser-Ser-Gly-Thr-Gly-Ile

site 4 gives a marginal response. In order to map the genes controlling the immune response to each one of these antigenic sites, studies were undertaken in numerous recombinant strains of mice. Only the results from the most critical strains used to map these genes will be discussed (Table 4). Recombinant strain B10.HTT has the high responder s allele in the K, A, B and J regions, low responder k alleles in the E, C and S regions. It was a high responder to sites 1, 2 and 5 whereas A.TL which has the s allele in the K and d allele in D and k in the middle was a low responder. This mapped the Ir genes to sites 1, 2 and 5 to the A, B or J regions. B10.S(9R) which is k in the J region was also a high responder to 1, 2 and 5, thereby narrowing down this mapping to either the A or the B region. B10.S(8R) narrowed down this mapping further within the I-A subregion. By similar analysis, immune response to site 4 and 5 also could be mapped to the I-A subregion. As shown previously, response to site 3 is controlled by non-H-2 genes. Figure 1 shows the map positions of the Ir-genes to each of the sites. These studies clearly showed that individual antigenic sites in a multi-determinant complex protein antigen are each controlled by unique Ir genes (8).

Table 4

Summary of Response to Myoglobin Antigenic Sites in Selected Recombinants

Strain	K	A	B	J	E	C	S	G	D	Response
B10	b	b	b	b	b	b	b	b	b	4
B10.D2	d	d	d	d	d	d	d	d	d	1,2,3,5
D2.GD	d	d	b	b	b	b	b	b	b	1,2,5
C3H.PL	d	d	d	d	d	d	k	k	k	1,2,3,5
B10.K	k	k	k	k	k	k	k	k	k	4
A.TL	s	k	k	k	k	k	k	k	d	4
B10.A	k	k	k	k	k	d	d	d	d	1,2,3,4
B10.A(2R)	k	k	k	k	k	d	d	?	b	1,2,4
B10.A(5R)	b	b	b	k	k	d	d	d	d	1,2,3,4
B10.S	s	s	s	s	s	s	s	s	s	1,2,5
B10.S(9R)	s	s	?	k	k	d	d	d	d	1,2,5
B10.HTT	s	s	s	s	k	k	k	k	d	1,2,5
B10.S(8R)	k	k	?	?	s	s	s	s	s	4

Table is from Okuda et al. (8)

Specificity of Anti-Myoglobin Antibodies in Mice and Other Species

The antigenic structure of Mb was initially determined (2) with early-course rabbit and goat antisera against Mb. The aforementioned studies on the genetic control of the immune response to Mb (5) and its synthetic antigenic sites (8) were carried out in congenic strains of mice by T-cell proliferation. Thus, using the Mb synthetic antigenic sites (determined with rabbit and goat antisera), we have shown (8) that these sites were effective in stimulating in vitro mouse T-cell proliferation and that the responses to the various antigenic sites in a protein are under separate genetic control. It appeared therefore that the molecular features recognized by B-cells are also recognized by T-cells. However, the conclusion was somewhat uncertain because the B-cell recognition sites (i.e. the antigenic sites) were not determined with mouse antibodies. This relates to the important question of whether the sites that are recognized as antigenic on a protein depend on the host species in which the antisera are raised. To answer this question, studies were carried out (9) in outbred and congenic strains of mice.

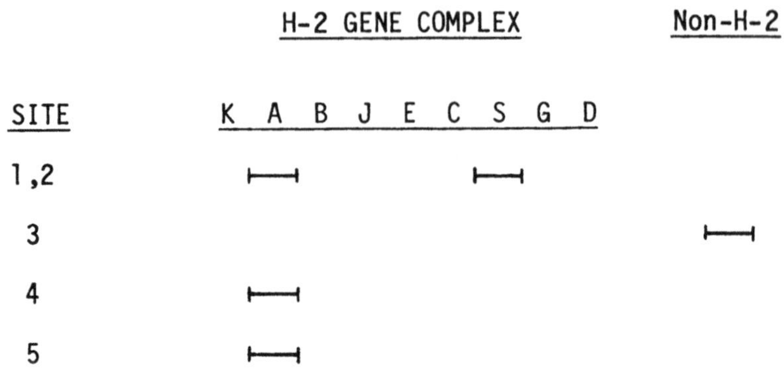

Figure 1. Ir gene mapping for individual antigenic sites of sperm-whale myoglobin. (From Okuda et al., ref. 8)

In the first studies (9), outbred mice were immunized with sperm-whale Mb (100 μg). Specific anti-Mb antibodies were isolated from each antiserum (100 μl) by adsorbing on Mb-sepharose as previously described for mouse antibodies against albumin (10). Specific antibodies were radioiodinated with ^{125}I. Immunoadsorbent titration studies using fixed amounts of ^{125}I-labelled specific antibodies permit the determination of whether all of the antibodies to a given antigen or antigenic site have been depleted (11). Each of the synthetic peptides corresponding to an antigenic site with rabbit and goat antisera was effective in binding mouse antibodies against Mb (Table 5). The reference Mb peptides which are non-antigenic with rabbit and goat antisera had no significant binding activity with mouse antibodies against Mb. The total amounts of antibodies bound by the five sites accounted for the entire (100%) antibody response for each outbred mouse to sperm-whale Mb. Only three examples are shown on Table 5.

Similar studies were performed in congenic strains of mice (9). An example of three of the strains is shown in Table 6. With congenic mice, lower antibody responses to some antigenic sites were observed. In A.SW and BALB/c the antibody to site 3 was quite low and with B10.S low amounts of antibodies were observed to both sites 3 and 4. It is noteworthy that the sites which generated low antibody responses in each of these congenic strains were the same sites which stimulated low in vitro T-cell

Table 5

Binding of Mouse (Outbred) ^{125}I-Labelled Antibodies With Immunoadsorbent of Mb and its Synthetic Sites

A fixed amount of ^{125}I-labelled antibodies against Mb from individual outbred mice were titrated with increasing amounts of Mb-Sepharose or peptide-Sepharose. For examples of such titrations see reference (11). The results below represent the maximum (plateau) amounts of antibodies that were bound by a given adsorbent. Values have been corrected for non-specific binding and are the average of six replicate analyses which varied ± 1.2% or less.

	Binding of ^{125}I-antibodies from outbred mice					
Antiserum:	MS2		MS3		MS5	
Immunoadsorbent	cpm bound	%bound	cpm bound	% bound	cpm bound	%bound
Mb	29,650	100	31,750	100	21,840	100
Site 1	4,320	14.6	5,080	16.0	3,150	14.4
Site 2	5,910	19.9	5,680	17.9	3,060	14.0
Site 3	8,250	27.8	5,210	16.4	4,590	21.0
Site 4	7,690	25.9	8,700	27.4	5,790	26.5
Site 5	3,460	11.7	7,020	22.1	5,200	23.8
Control peptide 1-6	0	0	0	0	<30	--
Control peptide 121-127	<40	--	<30	--	0	0
Total for sites 1-5	29,630	99.9	31,690	99.8	21,790	99.7

Table is from Twining et al. (9)

Table 6

Binding of Anti-Mb ^{125}I-Antibodies From Selected Congenic Mice With Immunoadsorbents of Mb and its Synthetic Sites

A fixed amount of ^{125}I-labelled antibodies against Mb from individual outbred mice were titrated with increasing amounts of Mb-Sepharose or peptide-Sepharose. For examples of such titrations see reference (11). The results below represent the maximum (plateau) amounts of antibodies that were bound by a given adsorbent. Values have been corrected for non-specific binding and are the average of six replicate analyses which varied ± 1.2% or less.

	Binding of ^{125}I-antibodies from congenic strains					
	ASW		BALB/c		B10.S	
Immunoadsorbent	cpm	%bound	cpm	%bound	cpm	%bound
Mb	23,910	100	27,630	100	20,090	100
Site 1	4,330	18.1	6,460	23.4	5,420	27.0
Site 2	6,410	26.8	7,430	26.9	6,230	31.0
Site 3	6,620	27.7	8,040	29.1	1,060	5.3
Site 4	597	2.5	980	3.5	920	4.6
Site 5	5,570	23.3	4,560	16.5	6,610	32.9
Control peptide 1-6	<30	--	<35	--	<30	--
Control peptide 121-127	0	--	0	--	<25	--
Total for sites 1-5	23,527	98.4	27,470	99.4	20,240	100.8
Sites Active in T-Cell Proliferation	1,2,3,5		1,2,3,5		1,2,5	

Table is from Twining et al. (9)

proliferation. At any rate, with the congenic strains the total amounts of antibodies directed to the five sites accounted in each case for 98-100% of the antibody response to Mb. It was concluded that (at least for Mb) the molecular features recognized by B-cells are also recognized by T-cells (9).

The question of whether antigenic recognition of a protein is dependent on the host species has been further extended (12) and we have shown that pig, cat and chicken antibodies against Mb recognize the same five antigenic sites that are recognized by rabbit, goat and mouse antisera. In other recent studies (13), the recognition of the Mb antigenic sites in rabbits from nine days up to a year have shown that the antigenic recognition of a protein (i.e. the regions that are recognized as antigenic) is not dependent upon the time antisera are obtained after the first immunization. These studies further confirm earlier conclusions (14-17) that the antigenic sites on a protein are structurally inherent in the protein.

Autoimmune Response to Myoglobin

The aforementioned findings on Mb, together with those on lysozyme (18), serum albumin (19-22) established, as just mentioned, that the antigenicity of protein antigenic sites is not host-dependent but is structurally inherent in their conformational uniqueness (14-17). Indeed, immunization with autologous Mb (16) or even the autologous serum albumin (21,22) readily generated auto-antibodies against the respective protein. Since the immune responses to sperm-whale Mb antigenic sites are under separate genetic control we have examined whether autoantigenic sites are also genetically controlled, and have determined T-cell participation in this autoimmune response. Mouse Mb was isolated from mice of the B10 background and its ability to invoke an autoimmune response in congenic mouse strains, also of the B10 background, was studied (23).

Mouse Mb from the heart muscle of mice (B10 background) was isolated, crystallized and the major chromatographic component was obtained by CM cellulose chromatography of the twice crystallized Mb by the procedure described elsewhere (24). The Mb thus obtained was homogeneous by disc gel electrophoresis. Mice were each injected in the tail with mouse Mb (100 μg) as a 1:1 (v/v) mixture of complete Freund's adjuvant and 0.01 M phosphate buffer, pH 7.2 containing 0.15 M sodium chloride. The T-cell lymphocyte proliferative response was studied by thymidine uptake on cultures of lumbar and inguinal lymph node cells obtained at ten days after immunization by a method similar to that of Corradin et al. (25).

The T-lymphocyte proliferative response was tested in several independent H-2 haplotypes (23). The results with the haplotypes

Table 7

Autoimmune Lymphocyte Proliferative Response to Mouse Mb (B10 Background) in Independent H-2 Haplotypes (B10 Background)

		[^{3}H] Thymidine Incorporation (CPM = Mean ± S.D.)[+]				
			Mouse Myoglobin*			PPD
Mouse Strains	Haplo-type	Medium or Lysozyme‡	10 μg/ml	150 μg/ml	250 μg/ml	50 μg/ml
B10	b	1012±258	1122±289 (1.11)	953±83 (0.94)	1199±87 (1.18)	45050±1959
B10.D2	d	1461±165	1725±161 (1.18)	1607±124 (1.10)	2450±351 (1.68)	85713±5134
B10.M	f	1719±249	1621±5 (0.94)	2900±56 (1.69)	2072±237 (1.21)	94215±6728
B10.WB	j	998±23	1489±64 (1.49)	1924±63 (1.93)	1878±438 (1.88)	79378±1618
B10.BR	k	2265±544	3946±313 (1.74)	3260±279 (1.44)	4560±225 (2.01)	84684±5774
B10.P	p	1251±154	1642±98 (1.31)	1693±66 (1.35)	2422±330 (1.94)	68644±4156
B10.Q	q	1415±117	1968±243 (1.39)	1714±19 (1.21)	2079±311 (1.47)	72481±4417
B10.S	s	1983±289	2245±117 (1.13)	17931±2204 (9.04)	19965±1758 (10.07)	107514±4449
B10.PL	u	1123±202	1606±68 (1.43)	1348±39 (1.20)	1515±140 (1.35)	86008±4327
B10.SM	v	1409±216	1847±251 (1.13)	1565±421 (1.11)	2104±13 (1.49)	58954±1003

[+]Values in parantheses represent stimulation indices.

‡Control cultures with 100 μg of lysozyme gave essentially the same level of incorporation as the medium.

*The non-responder strains still gave no response at 500 μg Mb/ml.

Table is from Yokota et al. (23)

Table 8

Mapping of the Ir Genes Controlling the Autoimmune Lymphocyte Proliferative Response to Mouse Myoglobin

Strains	H-2 Regions								$[^3H]$ Thymidine Incorporation (CPM = Mean ± S.D.)[+]		
	K	A	B	J	E	C	S	D	Medium or Lysozyme‡	Mouse Myoglobin 150 μg/ml	PPD 50 μg/ml
B10.TL	s	k	k	k	k	k	k	d	1768±608	2757±98 (1.56)	80466±2511
B10.S(7R)	s	s	s	s	s	s	s	d	2226±87	14527±780 (6.53)	106703±4361
B10.HTT	s	s	s	s	k	k	k	d	1976±678	12533±1193 (6.34)	104999±7189
B10.S(9R)	s	s	?	k	k	k	k	d	1050±130	9303±899 (8.86)	77492±7074
B10.S(8R)	k	k	?	?	s	s	s	s	1341±207	8925±821 (6.67)	78517±4504
B10.ASR2	k	k	k	k	k	d	d	s	1433±146	11175±367 (7.80)	116208±645
B10.DA	q	q	q	q	q	q	q	s	1621±191	2149±303 (1.33)	67726±6940

+Values in parantheses represent stimulation indices.

‡Control cultures with 100 μg of lysozyme gave essentially the same level of incorporation as the medium.

Table is from Yokota et al. (23).

of the B10 background are summarized in Table 7. The haplotypes $H\text{-}2^b$, $H\text{-}2^d$, $H\text{-}2^f$, $H\text{-}2^j$, $H\text{-}2^k$, $H\text{-}2^p$, $H\text{-}2^q$, $H\text{-}2^u$ and $H\text{-}2^v$ were either low responders or non-responders. However, the $H\text{-}2^s$ haplotype responded extremely well to challenge with mouse Mb. In order to map the genes controlling the autoimmune T-cell response to mouse Mb, selected recombinants (Table 8) were tested. The high response of B10.HTT and B10.S(9R) suggested the location of these genes in either the K or the I-A subregion. The low response of B10.TL ruled out the involvement of the K region and narrowed the mapping to the I-A subregion. The low response of B10.DA suggested that the H-2D end has no effect on autoimmune T-lymphocyte proliferative response. The high response of B10.S(8R) suggested a second Ir gene within the I-B and H-2D interval. However, B10.ASR-2 was surprisingly a high responder. There are two possible explanations. 1) The H-2D end effect is mediated by suppressor cells which act on the helper cells generated by Ir genes in the I-A subregion. This is a coupled interaction such that A^k and D^s generates response while A^q and D^s results in non-responsiveness. 2) The crossover points in B10.DA and B10.ASR-2 are at two different points in the S-D interval. This segment in B10.ASR-2 is derived from $H\text{-}2^s$ and maps the second Ir gene. The involvement of two separate Ir genes in the autoimmune response of Mb may suggest that autoantigenic sites, by analogy with the non-self antigenic sites (8) are each under separate genetic control. This is now being investigated.

Finally, it should be pointed out that the results we observed are true autoimmune responses and not generated by an altered Mb. In order to avoid the possibility that the Mb of some mouse strains may have suffered amino acid substitutions in its primary structure, we have isolated the Mb from mice with the same genetic background (B10). It will be entirely unlikely that the Mb will vary in primary structure among individual mice with a given genetic background.

STUDIES WITH MYOGLOBIN VARIANTS

Analysis of the Cross-Reactivity of Myoglobins in Terms of Site and Environmental Residue Substitutions

We have recently shown that the binding capacity of an antigenic site is altered not only by substitutions within site residues (i.e. the actual contact residues) but also by substitutions in neighboring residues which form the environment of the site (26,27,12). Furthermore, we have identified and calculated, from the X-ray coordinates of Mb, the distances of the nearest-neighbor residues (within 7.0 Å) which form the chemical environment for each antigenic site residue (27). Substitutions in the environmental residues can alter the expressed chemical characteristics of a site residue by changes in electron distribution, hydrogen bonding, steric effects and conformational readjustments. These factors have been treated in great detail elsewhere (26,27). It suffices here to mention that it has now been well established that evolutionary substitutions in site-neighboring residues can modify the characteristics and binding energy of the sites.

Recently, we have analyzed the immunochemical cross-reactions of 15 myoglobins, in terms of replacements in the antigenic sites and in the environmental residues (within 6.0 Å) of the sites, with sperm-whale Mb antisera raised in six different host species (12). The cross-reactivities are well rationalized by these effects. However, the complexity of the antigen (its multivalency) and the lack of knowledge of the amounts of antibodies directed to each sperm-whale Mb site in many of the antisera used, made it necessary to subject these analyses to additional rigorous testing. The relative amounts of antibodies directed to each antigenic site in the four inbred strains (Table 9) has been determined (9). Therefore, it will now be possible to evaluate with confidence the effects of amino acid substitutions in various myoglobins upon the reactivity of each site. The absence (or near absence) of antibodies to some antigenic sites should enable a less ambiguous analysis and dissection of substitution effects on the binding capacity of the reacting sites. We have recently carried out such analyses (28) and they will be summarized below.

Myoglobins from the skeletal muscle of bovine, sheep, goat, rabbit, finback whale and sperm whale and from the heart muscle of mice (outbred) and chicken were isolated and crystallized, and the major chromatographic components were obtained by CM-cellulose chromatography of the twice-crystallized myoglobins (6,24). Horse Mb from skeletal muscle was purchased from Sigma Chemical Co. (St. Louis, MO, U.S.A.) and further purified by CM-cellulose chromatography (24). Chimpanzee, cape-fox, dog, echidna and killer-whale myoglobins were purified by another procedure (29). Mice were each injected and boosted with sperm-whale Mb (100 μg) in PBS-complete Freund's adjuvant (1:1 v/v) by the same immunization schedule previously employed for raising rabbit antisera (30). Antisera were each a pool of four individual animals. Two inbred and two congenic strains (ASW, BALB/c, DBA/2 and B10.S) were studied. Specific anti-Mb antibodies were isolated from each antiserum by adsorbing on Mb-Sepharose and then, after radioiodination, the binding of these antibodies to the various myoglobins was determined by quantitative immunoadsorbent titration studies (11). The results summarized and discussed here represent the maximum (plateau) amounts of antibodies bound by a given immunoadsorbent.

Our previous analysis (9) enabled us to derive for each Mb variant the identity of the cross-reacting sites and the contribution of each to the overall cross-reaction of a given Mb. The

Table 9

Binding of ^{125}I-Antibodies Against Sperm-Whale Mb From Selected Congenic Mice With Immunoadsorbents of Various Myoglobins or The Synthetic Sites of Sperm-Whale Mb

A fixed amount of ^{125}I-labelled antibodies against sperm-whale Mb from inbred strains of mice (4 each) were titrated with increasing amounts of Mb-Sepharose or peptide-Sepharose. For examples of such titrations see (11). The results below represent the maximum (plateau) amounts of antibodies that were bound by a given adsorbent, relative to the amount bound in the plateau by sperm-whale Mb adsorbent as 100%. Values have been corrected for non-specific binding and are the average of six replicate analyses which varied ± 1.2% or less.

(a) Binding by site adsorbents+	Percentage of ^{125}I-Antibodies Bound From Congenic Strains			
Immunoadsorbent	ASW	BALB/c	DBA/2	B10.S
Site 1	18.1	23.4	28.0	27.0
Site 2	26.8	26.9	19.1	31.0
Site 3	27.7	29.1	24.2	5.3
Site 4	2.5	3.5	3.2	4.6
Site 5	23.3	16.5	23.6	32.9
Total for 5 sites	98.4	99.4	98.1	100.8

(b) Binding by Mb adsorbents

Mb adsorbents	Calc.	Found	Calc.	Found	Calc.	Found	Calc.	Found
Sperm Whale	100	100	100	100	100	100	100	100
Finback Whale	63.3	61.0	69.8	72.4	61.6	58.3	52.3	53.9
Killer Whale	58.7	56.9	63.9	64.3	54.6	53.2	45.5	46.4
Horse	52.0	50.3	57.3	55.9	47.5	46.1	37.8	39.2
Chimpanzee	44.5	47.2	49.0	49.6	40.7	39.2	35.3	34.8
Sheep	38.7	39.1	34.9	32.0	34.1	32.9	33.8	33.1
Bovine	38.7	37.2	34.9	32.5	34.1	32.3	33.8	33.0
Echidna	26.6	27.9	36.9	34.0	26.6	28.8	13.0	14.3
Rabbit	38.4	37.2	43.2	45.3	39.2	37.4	28.7	29.6
Cape Fox	32.0	34.4	34.3	32.5	29.8	26.6	18.5	18.4
Dog	32.0	33.5	34.3	33.3	29.8	26.4	18.5	18.8
Chicken	32.0	29.6	34.3	30.1	29.8	26.0	18.5	17.5
Calc. - Found Correlation Coeff.	0.996		0.997		0.997		0.999	

+Values are obtained from ref. (9).
Table is from Atassi et al. (28).

results are summarized in Table 10. From the expected level of cross-reaction of each site in a given Mb variant, and by knowing the relative amount of antibodies directed to each site on sperm-whale Mb, it will be possible to calculate the expected level of cross-reactivity for each Mb variant. Comparison of this with the value found experimentally will test the correctness or otherwise of the estimates previously made (12) on the effects of substitutions in the sites or within residues close to these sites (Table 10). For example, in finback whale Mb sites 1, 2 and 4 each retain about 75% of the reactivity of these sites in sperm-whale Mb, whereas site 3 retains full reactivity. Therefore, the expected cross-reaction of finback whale Mb with ASW antisera will be (18.1 x 0.75) + (26.8 x 0.75) + (27.7) + (2.4 x 0.75) = 63.3. The other expected reactivities were calculated by the same method (28).

It is significant that the values of cross-reactivities found experimentally for each Mb compared well with the values expected from the above considerations (28). These findings are very encouraging. However, it should be noted that these estimates are qualitative because, as previously pointed out (12), it is not possible to know the precise contribution of each site residue to the overall binding energy of the site and then the effect of a direct or an environmental substitution or a conformational readjustment on that binding energy. Nevertheless, operating within these constraints, it could be concluded (28) that the effects of substitutions within the sites and within the environmental residues of the sites constitute the major factors affecting the cross-reactions of homologous proteins.

Inter-Site Influences in T-Lymphocyte Proliferative Response to Myoglobin

The availability of congenic strains, in which the response to some antigenic sites is excluded by genetic control, and of Mb variants, in which the cross-reactivity alterations of some antigenic sites due to site and/or environmental substitutions have been determined, present a unique system of reduced complexity to analyze the molecular requirements for T-cell recognition of the Mb antigenic sites. It also affords a valuable tool to investigate inter-site influences in modulation of the overall T-cell recognition and response. Recently, we reported studies on the cross-reactions of ten myoglobins by proliferation of sperm-whale Mb-primed mouse T-cells (31). The cross-reactions were analyzed in terms of substitutions in the sites and in environmental residues of the sites.

Mice were each injected in the tail with sperm-whale Mb (100 µg) as a 1:1 (v/v) mixture of complete Freund's adjuvant and PBS. The procedure has been described (5). The T-lymphocyte

Table 10

Summary of the Reactivity of the Cross-Reacting Sites in Various Myoglobins With Sperm-Whale Mb Antibodies Expected on the Basis of Effects of Substitutions in the Antigenic Sites and in Their Environmental Residues

The effects of replacements within the antigenic sites are considered together with the effects of replacements in the neighboring residues. The expected reactivity of the site, therefore, takes into account the overall combined effects of these and the results can be expressed here only in relative qualitative terms (see Twining et al., ref. 12).

	REACTIVITY OF THE ANTIGENIC SITES[a]				
Myoglobin	Site 1	Site 2	Site 3	Site 4	Site 5
Sperm-Whale	F	F	F	F	F
Finback Whale	SD	SD	F	SD	0
Killer Whale	D	SD	F	SD	0
Horse	D	D	F	SD	0
Chimpanzee	D	D	SD	D	0
Sheep	GD	SD	D	GD	0
Beef	GD	SD	D	GD	0
Echidna	GD	0	SD	D	0
Rabbit	D	GD	SD	SD	0
Dog/Cape Fox	GD	GD	SD	0	0
Chicken	GD	GD	SD	0	0

[a]The letter notations are used as follows:

F: Indicates full reactivity (100%) is to be expected for the site.

SD: A slightly decreased level of binding activity (about 75%) for the site.

D: A decreased level of reactivity (about 50%) for the site.

GD: A greatly decreased level of reactivity (about 25%) for the site.

0: No reactivity is expected for the site.

Data presented in this table are from Twining et al. (12).

proliferative responses to Mb variants were studied by $[^3H]$-thymidine uptake on cultures of lumbar and inguinal lymph node cells, obtained at ten days after immunization, by a method similar to that of Corradin et al. (25) and recently described elsewhere (32). Immunizing and culture challenge doses were varied (1-200 μg) of Mb and the optimum immunizing and challenging doses were found to be 100 μg Mb per mouse and 50 μg/ml of culture, respectively.

The cross-reactions of various myoglobins by T-lymphocyte proliferative response were tested in four independent H-2 haplotypes and five recombinants. The antigenic sites of sperm-whale Mb recognized by these strains have been previously identified (8). Table 11 presents the findings with the independent haplotypes and serves to give an example of the T-cell proliferation data. Table 12 summarizes the stimulation indices with the independent haplotypes as well as the recombinant strains. It also gives the sites known to invoke a response in each mouse strain (8) as well as the cross-reactive sites in the Mb mutants and the changes in their cross-reaction as a result of amino acid substitutions within the sites and within environmental residues of the sites (12).

The $\underline{H\text{-}2^b}$ haplotype, which responds only to site 4, responded to sperm-whale, finback whale, killer whale and slightly to horse and rabbit Mb but showed no response to beef, sheep, cape fox, dog and chicken Mb. In finback whale, killer whale, horse and rabbit Mb, site 4 is only slightly altered (designated 4s in Table 12) by substitutions within site residues and/or in environmental residues (within 6.0 Å) of the sites (12). Therefore, the response of the T-cells from B10 is retained to this slightly altered site. On the other hand, site 4 in beef, sheep, cape fox, dog and chicken is either greatly altered or its reactivity is completely abolished. The $\underline{H\text{-}2^d}$ haplotype, which responds to sites 1, 2, 3 and 5 of sperm-whale Mb, gave high responses to sperm-whale, finback whale, killer whale, horse and rabbit Mb. It also responded very well, although at a lower level, to beef and sheep Mb, while no response was detected with cape fox, dog and chicken Mb. In these myoglobins, the reactivity of site 5 is destroyed by substitutions. The cross-reactions are therefore due to unaltered or perhaps altered forms of sites 1, 2 and 3. In finback whale Mb, the cross-reactions of sites 1 and 2 are slightly decreased, whereas site 3 is unaltered (i.e. sites are 1s, 2s, 3). In killer whale Mb, the cross-reaction of site 1 is decreased and those of sites 2 and 3 are slightly decreased and unaltered, respectively (i.e. sites are 1d, 2s, 3). In horse Mb the cross-reactions of the sites are: 1d, 2d, 3. From the T-cell proliferative cross-reaction of finback whale, killer whale and horse Mb, it is clear that sites 1 and 2 are important in $\underline{H\text{-}2^d}$ even as 1s, 2s, or 1d, 2d. Since the high cross-reaction of rabbit Mb is due to sites 1 and 3 in forms 1d, 3s, it is also clear that slight alteration of site 3 (i.e. 3s) will cause proliferation in conjunction with site 1 as 1d. Similarly, the good cross-reactive proliferation with beef and sheep Mb must be due to sites 2 and 3 as 2s, 3d. Clearly, then with this strain sites 1 and 2 are very important even as 1s or 1d and as 2s or 2d, and these in conjunction with site 3 (or 3s or 3d) stimulate T-cell proliferation. Altered site 3, by itself, did not stimulate proliferative response even as 3s. It appears, therefore, that the response to site 3 in the $\underline{H\text{-}2^d}$ haplotype requires help from the responses to sites 1 and/or 2.

Table 11

T-Lymphocyte Proliferative Responses to Various Myoglobins of Sperm-Whale-Primed T-Cells in Four Independent Haplotypes

Five mice of each strain were immunized subcutaneously in the tail with 100 μg of sperm-whale Mb in complete Freund's adjuvant. Ten days later, 4×10^5 draining lymph node cells were cultured with Mb variants at the final concentration of 50 μg/ml. After four days, the cultures were pulsed with $[^3H]$-thymidine for 24 hrs and the labelled cells were harvested and counted.

	$[^3H]$-Thymidine incorporation (cpm x 10^{-3} ± S.D.) in various haplotypes[+]			
	B10	B10.D2	B10.S	B10.M
Challenge	b	d	s	f
None (medium)	1.68 ± 0.19	1.77 ± 0.29	1.00 ± 0.09	1.18 ± 0.12
Lysozyme	1.59 ± 0.23	1.81 ± 0.22	0.99 ± 0.12	1.07 ± 0.23
Sperm Whale Mb	5.15 ± 0.60	63.22 ± 2.19	10.82 ± 0.60	11.47 ± 1.30
Finback Whale Mb	4.42 ± 0.32	60.26 ± 3.04	8.26 ± 0.54	7.84 ± 0.39
Killer Whale Mb	4.52 ± 0.25	57.91 ± 1.17	6.23 ± 0.24	6.23 ± 0.33
Horse Mb	3.85 ± 0.03	52.18 ± 1.56	2.21 ± 0.33	4.47 ± 0.44
Beef Mb	1.01 ± 0.05	25.17 ± 2.42	3.90 ± 0.27	5.51 ± 0.59
Sheep Mb	2.08 ± 0.13	26.35 ± 2.48	3.97 ± 0.26	6.33 ± 0.53
Cape Fox Mb	1.69 ± 0.13	1.39 ± 0.12	1.58 ± 0.13	1.42 ± 0.15
Dog Mb	1.04 ± 0.34	1.77 ± 0.13	1.26 ± 0.21	1.93 ± 0.18
Rabbit Mb	4.20 ± 0.12	61.04 ± 2.79	2.56 ± 0.18	2.47 ± 0.22
Chicken Mb	1.53 ± 0.04	2.42 ± 0.14	1.32 ± 0.03	1.51 ± 0.10
PPD	94.7 ± 2.25	84.34 ± 3.69	21.92 ± 0.35	79.05 ± 4.69

[+]Values represent the mean of triplicate analyses ± standard deviation.

Table is from Atassi et al. (31).

Table 12

Summary of T-Lymphocyte Proliferative Response to Various Myoglobins of Sperm-Whale Mb Primed T-Cells

Mice were injected with 100 μg of sperm-whale Mb with complete Freund's adjuvant in the tail subcutaneous tissue. After ten days, regional lymph node cells were collected and co-cultured with a Mb (50 μg) for three days. During the last 24 hrs, the cultures were pulsed with $[^3H]$-thymidine.

Strains	H-2 Regions	Sites recognized on S. Whale Mb +	S. Whale	F. Whale	K. Whale	Horse	Beef	Sheep	Cape Fox	Dog	Rabbit	Hen
B10	$K^bI^bS^bD^b$	4	3.1	2.6	2.7	2.3	0.6	1.2	1.0	0.6	2.5	0.9
B10.D2	$K^dI^dS^dD^d$	1,2,3,5	35.7	34.0	32.7	29.5	14.3	14.9	0.8	1.0	34.5	1.4
D2.GD	$K^dI^{d/b}S^bD^d$	1,2,5	34.3	50.5	51.4	41.1	44.7	44.7	1.4	0.8	42.7	1.7
B10.BDR1	$K^dI^dS^dD^b$	1,2,3,5	10.6	8.8	8.3	8.5	3.7	4.5	1.0	0.7	9.6	1.5
B10.S	$K^sI^sS^dD^s$	1,2,5	11.0	8.3	6.2	2.2	3.9	4.0	1.6	1.3	2.6	1.3
B10.S(26R)	$K^sI^sS^sD^b$	1,2,5	64.1	25.5	23.3	6.6	20.4	22.7	1.3	0.7	5.8	0.8
B10.HTT	$K^sI^{s/k}S^kD^d$	1,2,5	25.6	13.1	7.1	4.8	4.3	4.7	1.6	1.6	5.0	2.1
B10.M	$K^fI^fS^fD^f$	1,2,3	9.8	6.7	5.3	3.8	4.7	4.4	1.2	1.7	2.1	1.3
A.TFR4	$K^fI^fS^sD^d$	1,2,3	24.2	34.6	33.7	6.6	20.0	24.7	2.1	2.8	3.6	1.8
Site Reactivity in Various Myoglobins with Anti S.W. Mb‡			1,2,3, 4,5	1s,2s, 3, 4s	1d,2s, 3, 4s	1d,2d, 3, 4s	1g,2s, 3d,4g	1g,2s, 3d,4g	1g,2g 3s	1g,2g, 3s	1d,2g, 3s,4s	1g,2g 3s

+These data are from Okuda et al. (8) and Twining et al. (9).

‡The letter symbols following the site number are used to indicate the expected site reactivity in various myoglobins from site and/or environmental residue substitutions. S indicates a slightly decreased level of binding activity for the site; d, decreased level of reactivity for the site; g, greatly decreased level of reactivity for the site. For details, see Twining et al. (12).

Table is from Atassi et al. (31).

The recombinant D2.GD responds to sites 1, 2 and 5 only of sperm-whale Mb (8). The cross-reactive proliferative responses with finback whale, killer whale, horse, beef, sheep and rabbit Mb were very high and in fact were very significantly higher than the homologous response to sperm-whale Mb. On the other hand, cape fox, dog and chicken Mb did not stimulate a proliferative response. By analyses similar to those above based on consideration of the extent of the cross-reactivities of the five sites, expected from substitutions in the sites and in their environmental residues (12), certain conclusions can be deduced for the immune recognition of this strain. Site 5 in sperm-whale Mb exerts considerable suppressive influences on the overall response to sites 1 and 2, and when it is rendered unreactive by substitutions in Mb variants, a much higher response is obtained to sites 1 and 2 than in sperm-whale Mb. Sites 1 and 2 stimulate excellent responses in forms 1s, 2s (e.g. finback whale Mb) or even 1d, 2d (horse Mb). Furthermore, each site can act alone in either forms s or d. Thus the responses to beef and sheep Mb are triggered by site 2s, while the response to rabbit Mb is due to site 1d.

From the results with the recombinant B10.BDR1, which responds to sites 1, 2, 3 and 5 of sperm-whale Mb (8), it can be concluded that any slight alteration of site 3 even to 3s (as in cape fox, dog, chicken Mb) removes the response to this site. Sites 1 and 2 are important for the stimulation, even as 1s and 2s (finback whale Mb) or as 1d, 2d (horse Mb). Each alone can stimulate by itself (e.g. beef and sheep Mb responses due to site 2 as 2s), but site 1 is more important in stimulation than site 2, even as 1d (e.g. rabbit Mb).

The $\underline{H\text{-}2^s}$ haplotype and the recombinants B10.S(26R) and B10.HTT respond to sites 1, 2 and 5 of sperm-whale Mb (8,9). Site 5, which is rendered unreactive by site and environmental substitutions in all the present myoglobins (12), makes no contribution to the proliferative stimulation by the Mb variants. Sites 1 and 2 stimulate in the form 1s, 2s (finback whale Mb), but give only little stimulation in the form 1d, 2d (horse Mb). Either site by itself (e.g. 2s in beef and sheep Mb and 1d in rabbit Mb) stimulates less than the combination of the two. Neither site stimulates proliferation in the form 1g, 2g (cape fox, dog and chicken Mb).

The $\underline{H\text{-}2^f}$ haplotype and the recombinant A.TFR4 respond to sites 1, 2 and 3 of sperm-whale Mb (8). Any alteration of site 3, even to 3s renders it non-stimulatory, when presented by itself (cape fox, dog, chicken Mb). Sites 1 and 2 stimulate well in the form 1s, 2s (finback whale Mb) and less in the form 1d, 2d (horse Mb). Site 2 alone in the form 2s gives a good stimulation (e.g. beef and sheep Mb), which is lower than when it is present together with site 1, even when the latter is in the form of 1d (killer whale Mb). It is noteworthy, that in A.TFR4 the absence of a reactive form of site

5 leads to an enhancement of the response to sites 1, 2 and 3 in finback whale and killer whale Mb relative to sperm-whale Mb. It is concluded that site 5, although generating no proliferative response in A.TFR4, exerts a suppressive effect on the responses to sites 1, 2 and 3 in sperm-whale Mb.

In conclusion, the cross-reactions of Mb variants by proliferation of T-cells from strains that had been primed with sperm-whale Mb can be well rationalized in terms of substitutions in the antigenic sites and in environmental residues (within 7.0 Å) of the sites. The overall response is modulated by inter-site influences. The response to a site (or sites) can either be helped by the response to another as for example in the case of help between sites 1 and 2 and between these sites and site 3 in some of the strains discussed above. Alternatively, the response to or even the recognition of one site (e.g. site 5 in D2.GD and A.TFR4) could suppress the response to the whole molecule. These studies have thus enabled us to identify some of the molecular requirements for T-cell recognition of the Mb sites.

Gene Restriction of Antibody Responses to Antigenic Sites is Dependent on the Chemical Properties of the Sites

Previous studies, which have been outlined in the preceding sections, have shown that the chemical characteristics (and thus binding) of an antigenic site is influenced not only by substitutions in the sites, but also by those close (within 7.0 Å) to the sites (26,27,12). Thus in order to obtain a complete understanding of the chemical properties of an antigenic site, it is necessary to identify the environmental residues of the sites and the effects of any amino acid substitutions on their chemical properties. Substitutions within the sites taken together with substitutions in the environmental residues of the sites serve to evaluate the extent of the overall alteration of the site. These are summarized in Table 13. In atlantic bottlenosed dolphin Mb, sites 1, 2 and 5 are greatly altered, site 3 is only slightly altered, and site 4 suffers a medium alteration. In black sea dolphin Mb, sites 1, 2, 4 and 5 are drastically altered while site 3 is only slightly altered. In horse Mb, sites 1 and 2 suffer medium alterations, site 3 is unaltered, site 4 is slightly altered and site 5 is altered drastically. In badger Mb, sites 1, 2, 4 and 5 are drastically altered, while site 3 undergoes a slight alteration.

In addition to the foregoing intrinsic factors that influence and determine site recognition, there are factors that are extrinsic to the protein antigen. Of the extrinsic factors, those dictated by the genetic markers of the immune cells of the host animal are of relevance here. As we have already mentioned, the antigenic sites have been shown to be each under separate genetic control (8) so that in selected inbred strains of mice certain

Table 13

Summary of the Immune Responses to Myoglobin Variants in Different Strains of Mice

Selected H-2 Haplotypes and Recombinants	[a]Response to sperm-whale Mb: Responder Status	[a]Response to sperm-whale Mb: Site Responsiveness	[b]Response to Atlantic dolphin Mb. 1g, 2g, 3s, 4m, 5d	[b]Response to Black sea dolphin Mb. 1d, 2d, 3s, 4d, 5d	[b]Response to Horse Mb. 1m, 2m, 3u, 4s, 5d	[b]Response to Badger Mb. 1d, 2d, 3s 4d, 5d
$H\text{-}2^b$	Low	4	Low	High	Low	Low
$H\text{-}2^f$	High	1, 2, 3	Low	n.d.	n.d.	High
$H\text{-}2^k$	Low	4	Low	High	High	Low
$H\text{-}2^q$	Low	0	Low	High	High	Low
$H\text{-}2^r$	Low	0	High	n.d.	n.d.	High
$H\text{-}2^d$	High	1, 2, 3, 5	High	High	High	n.d.
B10.A(4R)	Low	4	Low	n.d.	n.d.	Low
B10.A(5R)	High	1, 2, 3, 4	Low	n.d.	n.d.	Low
B10.S(9R)	High	1, 2, 5	High	n.d.	n.d.	High
B10.HTT	High	1, 2, 5	High	n.d.	n.d.	n.d.
A.TH	High	1, 2, 5	High	n.d.	n.d.	High

a. Data from Okuda et al. (8)

b. Extent of overall alteration of antigenic sites 1-5 as compared to the corresponding sites of sperm-whale Mb. Abbreviations used: s, slightly altered; m, medium alteration; g, greatly altered; d, altered drastically; u, unaltered.

Table is from Young et al. (34).

strains mount significant T-cell (8) and antibody (4) responses only to certain sites. There appears to be no genetic control of antibody affinity (33). The availability of Mb variants in which the chemical alterations of their antigenic sites can be determined, and of inbred mouse strains whose responses to certain sites are known, will permit analysis of molecular factors that influence genetic control. Thus, in genetic control it is critical to know if the recognition of a site (or lack of it) is absolute regardless of the chemical properties of the site. For example, if a site on sperm-whale Mb is not recognized by a given strain, will that structural region be recognized by the same strain in another Mb if the chemical properties of the region are sufficiently different from sperm-whale Mb? The results reported recently (34) indicate that this is indeed the case. Table 13 summarizes some of the main points. It can be seen, for example, that the $\underline{H\text{-}2^b}$ haplotype, which responds only to site 4 and is thus a low responder to sperm-whale Mb, is a high responder to black sea dolphin but not to the other three myoglobins. The chemical alterations of the sites in black sea dolphin Mb effected a triggering of antibody response, while the alterations in the other myoglobins did not cause the appearance of a response. Similarly, the $\underline{H\text{-}2^k}$ haplotype, which responds only to site 4 in sperm-whale Mb, is able to respond to other sites in black sea dolphin Mb and in horse Mb. The most striking examples are those of the $\underline{H\text{-}2^q}$ and the $\underline{H\text{-}2^r}$ haplotypes which do not respond to any of the antigenic sites of sperm-whale Mb, but mount responses to the sites of black sea dolphin and horse Mb ($\underline{H\text{-}2^q}$) and of atlantic bottlenosed dolphin and badger Mb ($\underline{H\text{-}2^r}$). It is not known which of the five Mb antigenic sites become recognizable by various strains as a result of chemical alterations in different Mb variants. This must await the chemical synthesis of the regions and activity studies on the synthetic peptides.

In conclusion, the results suggest that genetic control of the responses to individual antigenic sites on a protein antigen is not only determined by the genetic constitution of the host species but also by the chemical properties of the individual sites. It appears that the H-2 subregions mapping the responses to given antigenic sites can also recognize other sites, which were previously unrecognizable in a homologous protein, if their chemical properties are suitably altered. For example, it is possible that an antigenic site in one myoglobin could act as a suppressor site, and that the corresponding chemically altered antigenic site in another Mb variant could act as a helper site. So, whereas the structural locations of the antigenic sites are constant within a given family of proteins, the same or other H-2 subregions may control responses to different sites on different proteins within this family.

ACKNOWLEDGEMENTS

This work was supported by grants AM-18920 from the National Institute of Arthritis and Metabolic Diseases, AI-18657 from the

National Institute of Allergy and Infectious Diseases and CA-24473 from the National Cancer Institute, National Institutes of Health, U.S. Public Health Service.

REFERENCES

1. D.C. Shreffler and C.S. David, The H-2 major histocompatibility complex and the immune response region. Genetic variation, function and organization. Adv. Immunol. 20:125 (1975).
2. M.Z. Atassi, Antigenic structure of myoglobin: The complete immunochemical anatomy of a protein and conclusions relating to antigenic structures of proteins. Immunochemistry 12:423 (1975).
3. A.B. Edmundson, Amino acid sequence of sperm-whale myoglobin. Nature 205:883 (1965).
4. J.C. Kendrew, H.C. Watson, B.E. Strandberg, R.E. Dickerson, D.C. Phillips and V.C. Shore, A partial determination of X-ray methods and its correlation with chemical data. Nature (Lond.) 190:666 (1961).
5. K. Okuda, P.R. Christadoss, S. Twining, M.Z. Atassi and C.S. David, Genetic control of the immune response to sperm-whale myoglobin in mice. I. T-lymphocyte proliferative response under H-2-linked Ir gene control. J. Immunol. 121:866 (1978).
6. M.Z. Atassi, Properties of components of myoglobin of the sperm whale. Nature (Lond.) 202:496 (1964).
7. R.H. Schwartz, L. Jackson and W.E. Paul, T-lymphocyte enriched murine peritoneal exudate cells. I. A reliable assay for antigen-induced T-lymphocyte proliferation. J. Immunol. 115:1330 (1975).
8. K. Okuda, S.S. Twining, C.S. David and M.Z. Atassi, Genetic control of the immune response to sperm-whale myoglobin in mice. II. T-lymphocyte proliferative response to the synthetic antigenic sites. J. Immunol. 123:182 (1979).
9. S.S. Twining, C.S. David and M.Z. Atassi, Genetic control of the immune response to myoglobin. IV. Mouse antibodies in outbred and congenic strains against sperm-whale myoglobin recognize the same antigenic sites that are recognized by antibodies raised in other species. Mol. Immunol. 18:447 (1981).
10. S. Sakata and M.Z. Atassi, Immune recognition of serum albumin. XI. Mouse antibodies against bovine serum albumin recognize the same antigenic sites that are recognized by rabbit antibodies. Biochim. Biophys. Acta 625:159 (1980).
11. S.S. Twining and M.Z. Atassi, Use of immunoadsorbents for the study of antibody binding to sperm-whale myoglobin and its synthetic antigenic sites. J. Immunol. Methods 30:139 (1979).

12. S.S. Twining, H. Lehmann and M.Z. Atassi, The antibody response to myoglobin is independent of the immunized species: Analysis in terms of replacements in the antigenic sites and in environmental residues of the cross-reactions of fifteen myoglobins with sperm-whale myoglobin antisera raised in different species. Biochem. J. 191:681 (1980).
13. S.S. Twining, H. Lehmann and M.Z. Atassi, Time-dependence study of the antibody response to sperm-whale myoglobin: Recognition of the antigenic sites is unaltered over an extended period of immunization. Mol. Immunol. 18:473 (1981).
14. A.L. Kazim and M.Z. Atassi, Prediction and confirmation by synthesis of two antigenic sites in human haemoglobin by extrapolation from the known antigenic structure of sperm-whale myoglobin. Biochem. J. 167:275 (1977).
15. A.L. Kazim and M.Z. Atassi, Antibodies against protein antigenic sites that are identical in the homologous protein of the immunized animal: Autoreactivity in rabbits of antibodies to sperm-whale myoglobin. Biochim. Biophys. Acta 494:277 (1977).
16. A.L. Kazim and M.Z. Atassi, Production of autoantibodies by immunization with rabbit myoglobin. Immunochemistry 15:67 (1978).
17. M.Z. Atassi and A.L. Kazim, First consequences of the determination of the entire antigenic structure of sperm-whale myoglobin in "Immunobiology of Proteins and Peptides" (Atassi, M.Z. and Stavitsky, A.B., Eds.) Vol. 1, pp. 19-40, Plenum Press, New York, NY.
18. M.Z. Atassi, Precise determination of the entire antigenic structure of lysozyme: Molecular features of protein antigenic structures and potential of "surface-simulation" synthesis, a powerful new concept for protein binding sites. Immunochemistry 15:909 (1978).
19. M.Z. Atassi, S. Sakata and A.L. Kazim, Immunochemistry of serum albumin. IX. Localization and verification by synthesis of five antigenic sites of bovine serum albumin. Biochem. J. 179:327 (1979).
20. S. Sakata and M.Z. Atassi, Immunochemistry of serum albumin. X. Five major antigenic sites of human serum albumin are extrapolated from bovine albumin and confirmed by synthetic peptides. Mol. Immunol. 17:139 (1980).
21. S. Sakata and M.Z. Atassi, Autorecognition of serum albumin. Fed. Proc. 39, Abstr. 3428 (1980).
22. S. Sakata and M.Z. Atassi, Immune recognition of serum albumin. 13. Autoreactivity with rabbit serum albumin of rabbit antibodies against bovine or human serum albumins and autoimmune recognition of rabbit serum albumin. Mol. Immunol. 18:961 (1981).

23. S. Yokota, C.S. David and M.Z. Atassi, Genetic control of the immune response to myoglobin. III. Autoimmune T-lymphocyte proliferative response to mouse myoglobin. Mol. Immunol. 17:1079 (1980).
24. M.Z. Atassi, Immunochemistry of sperm-whale myoglobin. VI. Preparation and conformational analysis of eight mammalian myoglobins. Biochim. Biophys Acta 221:612 (1970).
25. G. Corradin, H.M. Etlinger and J.M. Chiller, Lymphocyte specificity to protein antigens. I. Characterization of the antigen induced in vitro T-cell dependent proliferative response with lymph node cells from primed mice. J. Immunol. 119:1048 (1977).
26. M.Z. Atassi and A.L. Kazim, Distance calculation of residues neighboring to lysozyme antigenic sites: Site neighboring residues whose evolutionary substitution can modify the characteristics and binding energy of the sites. Biochem. J. 187:163 (1980).
27. A.L. Kazim and M.Z. Atassi, Nearest neighbor analysis of myoglobin antigenic sites: Nearest neighbor residues whose replacement can alter the environment of binding site residue(s) and thus change their characteristics and binding ability. Biochem. J. 191:673 (1980).
28. M.Z. Atassi, S.S. Twining, H. Lehmann and C.S. David, Genetic control of the immune response to myoglobin. V. Analysis of the cross-reactivity of 12 myoglobins with sperm-whale myoglobin antisera of inbred mouse strains in terms of substitutions in the antigenic sites and in the environmental residues of the sites. Immunol. Commun. 10:359 (1981).
29. A.E. Romero-Herrera, H. Lehmann and O. Castillo, The myoglobin of primates. VIII. N,Y-cticebus cougane (Slow Loris). Biochim. Biophys Acta 420:387 (1976).
30. M.Z. Atassi, Periodate oxidation of sperm-whale myoglobin and the role of the methionines in the antigen-antibody reaction. Biochem. J. 102:478 (1967).
31. M.Z. Atassi, S. Yokota, S.S. Twining, H. Lehmann and C.S. David, Genetic control of the immune response to myoglobin. VI. Inter-site influences in T-lymphocyte proliferative response from analysis of cross-reactions of ten myoglobins in terms of substitutions in the antigenic sites and in environmental residues of the sites. Mol. Immunol. 18:945 (1981).
32. K. Okuda, S. Sakata, M.Z. Atassi and C.S. David, Genetic control of the immune response to hen's egg-white lysozyme in mice. I. Antibody and T-lymphocyte proliferative responses to the native protein. J. Immunogen. 6:447 (1979).
33. C.R. Young, G.P. O'Connor and M.Z. Atassi, Genetic control of the immune response to myoglobin. VIII. Extent of the genetic control of antibody affinity. J. Immunogen. 8:387 (1981).

34. C.R. Young, G.P. O'Connor and M.Z. Atassi, Genetic control of the immune response to myoglobin. VII. Antibody responses to myoglobin variants reveal that gene restriction of the antibody responses to myoglobin antigenic sites is dependent on the chemical properties of the sites. Immunol. Commun. 10:483 (1981).

7

GENETIC CONTROL OF THE IMMUNE RESPONSE TO HEMOGLOBIN AND ITS SUBUNITS

Christopher J. Krco, A. Latif Kazim, M. Zouhair Atassi, and Chella S. David

Department of Immunology
Mayo Clinic
Rochester, MN 55905

SUMMARY

Human adult hemoglobin (Hb) was selected as a model oligomeric protein for investigations into Ir gene function. Separate genetic control of the α- and β-subunits of Hb was demonstrated. The relevant Ir genes were mapped to the I-A subregion and in particular to the A_β gene for the α-chain of Hb. $H\text{-}2^b$ and $H\text{-}2^d$ mice are responders to the α-subunit and $H\text{-}2^d$, $H\text{-}2^f$ and $H\text{-}2^s$ mice are responsive to the β-subunit. The subunits are not cross-reactive in T-cell assays. Non-H-2 linked genes are also important in Hb-specific immunity. There is a requirement for $Ly\text{-}1^+$ T-cells for optimal responses in vitro.

INTRODUCTION

That the immune response to foreign proteins and synthetic peptides is under control of H-2 linked genes has been amply documented (1). Many studies of Ir gene control have involved the use of synthetic polymers of a limited number of amino acids. However, more recently native protein molecules such as staphylcoccal nuclease (2), myoglobin (3-9), cytochrome c (10) and lysozyme (11,12) have been used as model antigens. The foregoing studies have employed monomeric proteins (i.e., composed of a single polypeptide chain under normal physiological conditions). With only a few exceptions, lactate dehydrogenase (13,14), insulin (15) and thyroglobulin (16,17), little work has been done on oligomeric proteins. However, thyroglobulin is a very large molecule which is not well characterized; lactic dehydrogenase has identical subunits as does the insulin dimer. Thus, relatively little information is available concerning the immunogenetics of responses to oligomeric proteins. For example, it is unknown if during an immune response to an oligomeric protein whether the different subunits will exert a cooperative or regulating effect upon the response to one another. As a model oligomeric protein in this regard we have selected human hemoglobin (Hb) to serve as an antigenic probe and is part of our effort to determine the antibody and cellular recognition of Hb (18,19). The results of such investigations are reported here.

MATERIALS AND METHODS

Details of antigen preparation, immunization and culture conditions have previously been published (20).

RESULTS

To determine if H-2 linked genes are associated with immune responsiveness to native hemoglobin (Hb), mice of several independent haplotypes were immunized using 100 μg of Hb emulsified in complete Freund's adjuvant (20). The draining lymph node cells were removed and challenged in vitro with Hb and its subunits (Table 1). Mice of the $H\text{-}2^b$ (B10) and $H\text{-}2^d$ (B10.D2) haplotypes are high responders to Hb (Δcpm of approximately 47,000) and its α-chain subunit (Δcpm of 39,000 and 53,000 respectively). However, only B10.D2 cells remain responsive to challenge with the β-subunit (Δcpm 38,000 vs. 5,000). In contrast mice of other independent haplotypes ($H\text{-}2^j$, $H\text{-}2^k$, $H\text{-}2^q$, $H\text{-}2^r$, $H\text{-}2^s$) were classified as low responders to both Hb and its subunits (Δcpm less than 10,000). These results are consistent with the existence of H-2 linked Ir genes.

In order to map the Ir genes within the mouse major histocompatibility complex (MHC) a series of recombinant mouse strains (Table 2) were tested. The high responsiveness of B10 to Hb (Δcpm 48,000) and to the α-chain subunit (Δcpm 39,000) and the

Table 1
H-2 LINKED GENES DETERMINE IMMUNE RESPONSIVENESS TO HEMOGLOBIN

		Δcpm ± SEM x 10^{-3}		
Mouse Strain	Haplotype	Hemoglobin	α-chain	β-chain
B10	b	48.0 ± 1.0	39.0 ± 1.0	5.0 ± 0.8
B10.D2	d	46.0 ± 5.0	53.0 ± 4.0	38.0 ± 0.1
B10.WB	j	7.0 ± 1.0	5.0 ± 0.2	10.0 ± 1.0
B10.K	k	3.0 ± 0.2	1.0 ± 0.1	4.0 ± 1.0
B10.Q	q	2.0 ± 0.1	1.0 ± 0.3	1.0 ± 0.2
B10.RIII	r	4.0 ± 0.8	2.0 ± 0.4	0.1 ± 0.2
B10.S	s	8.0 ± 0.6	5.0 ± 0.6	8.0 ± 1.0

Table 2
I-A SUBREGION GENES ARE ASSOCIATED WITH RESPONSIVENESS TO HEMOGLOBIN AND ITS SUBUNITS

Mouse Strain	MHC Regions and Alleles[a] K A B J E C S D	Δ cpm ± SEM x 10^{-3} Hemoglobin	α-chain	β-chain
B10	b b b b b b b b	48.0 ± 1.0	39.0 ± 1.0	5.0 ± 0.8
B10.A	k k k k k d d d	10.0 ± 1.0	-2.0 ± 0.1	20.0 ± 1.0
B10.OL	d d d d d d k k	43.0 ± 2.0	37.0 ± 2.0	18.0 ± 2.0
B10.AKM	k k k k k k k q	2.0 ± 0.8	2.0 ± 0.4	2.0 ± 0.4
B10.A(4R)	k k b b b b b b	3.0 ± 1.0	6.0 ± 1.0	4.0 ± 0.5
B10.MBR	b k k k k k k q	1.0 ± 0.4	0.7 ± 0.1	1.0 ± 0.2

[a]Alleles associated with responsiveness are underlined.

low responsiveness of B10.A ($K^kA^kB^kJ^kE^kC^dS^dD^d$) to both antigens (Δcpm less than 11,000) maps the relevant Ir genes to the left of the I-C subregion. This intrepretation is supported by the responsiveness of B10.OL ($K^dA^dB^dJ^dE^dC^dS^kD^k$) to both Hb (Δcpm 43,000) and α-chain (Δcpm 37,000). The low responses obtained when challenging both B10.A(4R) ($K^kA^kB^bJ^bE^bC^bS^bD^b$) and B10.MBR ($K^bA^kB^kJ^kC^kS^kD^q$) to Hb or α-chain (Δcpm less than 7,000) maps Hb- and α-chain-specific Ir genes to the I-A subregion of the MHC. The responsiveness of B10.OL and B10.A to challenge using the β-subunit of Hb (Δcpm of approximately 19,000) maps two β-chain-specific Ir genes: one to the left of the I-C subregion and a second to the H-2D end of the MHC.

To definitively map the Ir genes to the β-chain, another series of recombinant mice were tested (Table 3). The low responsiveness of B10.S ($K^sA^sB^sJ^sE^sC^sS^sD^s$) and the high responsiveness of B10.S(9R) ($K^sA^sB^?J^kE^kC^dS^dD^d$) and B10.S(7R) ($K^sA^sB^sJ^sE^sC^sS^sD^d$) (Δcpm of 34,000 and 46,000 respectively) demonstrate that H-2D end genes influence the immune response to Hb and the β-subunit. These strains remain unresponsive to the α-chain subunit (Δcpm less than 3,000) thus demonstrating the β-chain specificity of the H-2D end genes. Interestingly, the replacement of the D^s allele with either a D^r [B10.RIII(20R); $K^sA^sB^sJ^sE^sC^sS^sD^r$] or a D^b [B10.S(26R); $K^sA^sB^sJ^sE^sC^sS^sD^b$] allele, both normally associated with β-chain unresponsiveness, will render Hb-primed I^s haplotype mice responsive to the β-subunit. The observation that these H-2D end recombinant mice are unresponsive to α-chain and only half as responsive (Table 3) to intact Hb molecules suggests that the D^s allele, in conjunction with I^s Ir genes, recognizes α-chain determinants in a suppressive fashion. In the absence of the suppressive D^s influence, I^s haplotype mice can respond normally.

Table 3
H-2D END GENES ARE ASSOCIATED WITH RESPONSIVENESS TO HEMOGLOBIN AND ITS β-CHAIN SUBUNIT

Mouse Strain	MHC Regions and Alleles[a] K A B J E C S D	Δ cpm ± SEM x 10^{-3} Hemoglobin	α-chain	β-chain
B10.S	s s s s s s s s	8.0 ± 0.6	5.0 ± 0.6	8.0 ± 1.0
B10.S(9R)	s s ? k k d d d	15.0 ± 3.0	1.0 ± 0.1	34.0 ± 1.0
B10.S(7R)	s s s s s s s d	43.0 ± 2.0	2.0 ± 0.5	46.0 ± 4.0
B10.T(6R)	q q q q q q q d	3.0 ± 0.1	1.0 ± 0.1	2.0 ± 0.1
B10.S(20R)	s s s s s s s r	29.0 ± 3.0	9.0 ± 0.03	51.0 ± 0.3
B10.S(26R)	s s s s s s s b	8.0 ± 0.4	-2.0 ± 0.4	17.0 ± 1.0

[a]Alleles associated with responsiveness are underlined.

Since the mice utilized in the experiments discussed to this point were immunized with Hb (hence both subunits simultaneously) the possibility exists that Hb subunits might negatively influence the immune response to one another. This situation may be especially relevant in H-2^s haplotype mice. Experiments were conducted therefore, using mice that had been primed to each subunit separately (21).

It was observed (Table 4) that only H-2^b (B10) and H-2^d (B10.D2) haplotype mice are responsive to immunization and challenge to the α-chain of Hb (Δcpm of approximately 30,000). All other nine independent haplotype mice are low responders (Δ cpm less than 9,000). The relevant Ir genes were mapped to the I-A subregion (Table 5) based upon the low responsiveness of B10.A(4R) ($K^kA^kB^bJ^bE^bC^bS^bD^b$) and B10.MBR ($K^bA^kB^kJ^kE^kC^kS^kD^q$) (Δ cpm less than 4,000). The responsiveness (Δcpm 25,000) of B10.A(5R) ($K^bA^bB^bJ^kE^kC^dS^dD^d$) suggests that genes within the I-J and I-E subregions are not crucial to α-chain immunity. In order to map the Ir genes within the I-A subregion advantage was taken on the I-A mutant mouse strain B6.C-H-2^{bm12} (22). The low responsiveness of bm12 mice (Δcpm -100) when compared to the responsiveness of parental B6/Kh and reference B10 mice (Δcpm 33,000) maps the α-chain-specific Ir genes to the A_β gene and suggests that the Ia.8 specificity is intimately associated with α-chain responsiveness (22).

Table 4
STRAIN DISTRIBUTION OF T-CELL PROLIFERATIVE RESPONSES TO α-CHAIN OF HEMOGLOBIN

Mouse Strain	Haplotype	Δcpm $\pm$ SEM x 10^{-3}
B10	b	28.0 ± 2.0
B10.D2	d	31.0 ± 0.6
B10.M	f	1.0 ± 0.1
B10.WB	j	5.0 ± 0.1
B10.K	k	2.0 ± 0.2
B10.F	p	0.4 ± 0.1
B10.G	q	0.3 ± 0.1
B10.RIII	r	8.0 ± 0.3
B10.S	s	3.0 ± 0.6
B10.PL	u	5.0 ± 1.0
B10.SM	v	5.0 ± 0.4

Table 5
I-A (A_β) SUBREGION GENES DETERMINE RESPONSIVENESS TO THE α-CHAIN SUBUNIT OF HEMOGLOBIN

	MHC Regions and Alleles[a]								
Mouse Strain	K	A	B	J	E	C	S	D	Δcpm $\pm$ SEM x 10^{-3}
B10.A(5R)	$\underline{b}$	$\underline{b}$	$\underline{b}$	k	k	$\underline{d}$	$\underline{d}$	$\underline{d}$	25.0 ± 0.5
B10.A(4R)	k	k	$\underline{b}$	$\underline{b}$	$\underline{b}$	$\underline{b}$	$\underline{b}$	$\underline{b}$	1.5 ± 0.1
B10.MBR	$\underline{b}$	k	k	k	k	k	k	q	3.0 ± 0.2
B10	$\underline{b}$	$\underline{b}$	$\underline{b}$	$\underline{b}$	$\underline{b}$	$\underline{b}$	$\underline{b}$	$\underline{b}$	33.0 ± 1.0
B6/kh	$\underline{b}$	$\underline{b}$	$\underline{b}$	$\underline{b}$	$\underline{b}$	$\underline{b}$	$\underline{b}$	$\underline{b}$	33.0 ± 0.3
B6.C-H-2^{bm12}	$\underline{b}$	$\underline{b}$*	$\underline{b}$	$\underline{b}$	$\underline{b}$	$\underline{b}$	$\underline{b}$	$\underline{b}$	-0.1 ± 0.03

[a]Alleles associated with responsiveness are underlined.

In parallel experiments (Table 6) mice were primed to the β-subunit and challenged in vitro (14). In contrast to the situation with α-chain (Table 4) three patterns of responsiveness (high, intermediate, low) exist with regards to the β-subunit of Hb. Mice of the H-2^d (B10.D2), H-2^f (B10.M) and H-2^s (B10.S) haplotypes are classified as high responders (Δcpm 59,000, 39,000, and 62,000 respectively). H-2^k (B10.K), H-2^p (B10.F), H-2^r (B10.RIII), H-2^v (B10.SM) and H-2^u (B10.PL) are intermediate responders (Δcpm of 14,000-18,000). Mice of the H-2^b (B10), H-2^j (B10.WB) and H-2^q (B10.G) are low responders (Δcpm less than 10,000).

Table 6
STRAIN DISTRIBUTION OF T-CELL PROLIFERATIVE RESPONSES TO THE β-CHAIN SUBUNIT OF HEMOGLOBIN

Mouse Strain	Haplotype	Δcpm ± SEM x 10^{-3}
B10.D2	b	59.0 ± 3.0
B10.M	f	39.0 ± 2.0
B10.S	s	62.0 ± 0.1
B10.K	k	16.0 ± 3.0
B10.F	p	18.0 ± 2.0
B10.RIII	r	15.0 ± 1.0
B10.SM	v	14.0 ± 1.0
B10.PL	u	14.0 ± 1.0
B10.WB	j	7.0 ± 1.0
B10.G	q	9.0 ± 2.0
B10	b	2.4 ± 0.6

The β-chain responsiveness of B10.S(26R) ($K^sA^sB^sJ^sE^sC^sS^sD^b$) maps the Ir genes to the left of the S subregion (Table 7). The low responsiveness (Δcpm 7,000) of B10.S(8R) ($K^kA^kB^kJ^kE^sC^sS^sD^s$) in contrast to the responsiveness (Δcpm of 42,000) of B10.HTT ($K^sA^sB^sJ^sE^kC^kS^dD^d$) maps the β-chain Ir genes to the left of the I-J subregion. The Ir genes are pinpointed to the I-A subregion by the low response (Δcpm 9,000) of B10.TL ($K^sA^kB^kJ^kE^kC^kS^dD^d$) when compared to the strong response elicited by B10.S(9R) ($K^sA^sB^?J^kE^kC^dS^dD^d$) mice.

Table 7
I-A SUBREGION GENES ARE ASSOCIATED WITH RESPONSIVENESS TO THE β-CHAIN SUBUNIT OF HEMOGLOBIN

	MHC Regions and Alleles[a]								
Mouse Strain	K	A	B	J	E	C	S	D	Δcpm ± SEM x 10^{-3}
B10.S(26R)	s	s	s	s	s	s	s	b	48.0 ± 8.0
B10.S(8R)	k	k	k	k	s	s	s	s	7.0 ± 0.2
B10.HTT	k	k	k	k	s	s	s	d	42.0 ± 0.5
B10.S(9R)	s	s	?	k	k	d	d	d	57.0 ± 1.4
B10.TL	s	k	k	k	k	k	k	d	9.0 ± 0.4

[a]Alleles associated with responsiveness are underlined.

The patterns of immune responsiveness to the Hb subunits was indicative of separate genetic control. To formally address this question mice of the four haplotypes, $H\text{-}2^b$, $H\text{-}2^d$, $H\text{-}2^f$, $H\text{-}2^s$, associated to responsiveness to Hb subunits were immunized and challenged with either subunit (Table 8). The data from such experiments reveals that three phenotypes with respect to Hb subunit responsiveness exist. Phenotype I is associated with responsiveness to the α-chain and is exemplified by B10 ($H\text{-}2^b$) mice which are α-chain responders (Δcpm 25,000) but which are unresponsive to β-chain (Δcpm 400). Phenotype II mice exhibit the reciprocal pattern. Both B10.M ($H\text{-}2^f$) and B10.S ($H\text{-}2^s$) mice respond to β-chain (Δcpm of 35,000 and 83,000 respectively) while maintaining unresponsiveness to α-chain determinants (Δcpm of approximately 4,500). Phenotype III is characterized by responsiveness to either subunit (B10.D2 mice). When primed with α-chain (Δcpm 20,000) but will not cross-react with β-chain (Δcpm 1,000). Conversely, B10.D2 cells will not respond to α-chain (Δcpm 2,000) if the cells had been previously primed to β-chain (Δcpm 66,000). Thus, the α- and β-subunits do not cross-react in T-cell assays.

Table 8
α- AND β-SUBUNITS OF HEMOGLOBIN ARE NOT CROSS-REACTIVE IN T-CELL PROLIFERATION ASSAYS

Phenotype	Prototype Strain	Priming Antigen	Δcpm $\pm$ SEM x 10^{-3} α-Chain	β-Chain
I	B10	α-Chain	25.0 ± 3.0	0.4 ± 0.02
II	B10.M	β-Chain	4.0 ± 0.1	35.0 ± 2.0
	B10.S		5.0 ± 1.0	83.0 ± 2.0
III	B10.D2	α-Chain	20.0 ± 4.0	1.0 ± 0.3
		β-Chain	2.0 ± 0.6	66.0 ± 3.0

As has been demonstrated for other protein antigens (5,16, 23-28) there exists non-H-2 genes that affect immunity to Hb (Table 9). While B10 and A.BY mice (both $H\text{-}2^b$) respond to α-chain (Δcpm of approximately 21,500), C3H.SW mice are low responders (Δcpm 4,000). Balb/c mice ($H\text{-}2^d$) respond half as well to α-chain and are unresponsive to β-chain when compared to B10.D2 mice. These results demonstrate that there are non-H-2 linked genes which affect Hb subunit immunity.

Table 9
NON-H-2 GENES INFLUENCE RESPONSIVENESS TO EITHER SUBUNIT OF HEMOGLOBIN

Mouse Strain	Haplotype	Priming Antigen	Δcpm $\pm$ SEM x 10^{-3}
B10	b	α-chain	21.0 ± 1.0
A.BY	b		22.0 ± 1.0
C3H.SW	b		4.0 ± 0.5
B10.D2	d		33.0 ± 4.0
BALB/C	d		17.0 ± 2.0
B10.D2	d	β-chain	102.0 ± 9.0
BALB/c	d		1.0 ± 0.9
B10.S	s		38.0 ± 0.5
A.SW	s		29.0 ± 0.6

The in vitro response to both α- and β-subunits are dependent upon the presence of Ly-1+ T-cells (Table 10). Pretreatment of α-chain primed B10 or β-chain primed B10.S cells with anti-Ly-1 plus complement (29) will inhibit (88% and 94% respectively) subsequent antigen specific T-cell proliferative responses. Pretreatment of either cell source with anti-Ly-2 plus complement has only minimal effects.

Table 10
Ly-1+ T-CELLS ARE REQUIRED FOR RESPONSIVENESS TO EITHER SUBUNIT OF HEMOGLOBIN

Mouse Strain	Treatment	Δcpm ± SEM x 10^{-3}
	--	20.0 ± 2.0
B10	αLy-1.2+C'	3.0 ± 0.5
	αLy-2.2+C'	22.0 ± 1.0
	--	25.0 ± 1.5
	αLy-1.1+C'	20.0 ± 1.0
B10.S		
	αLy-1.2+C'	2.0 ± 0.5
	αLy-2.2+C'	19.0 ± 0.5

DISCUSSION

The data reported here have permitted the identification of Ir genes that determine responsiveness to Hb and its constituent subunits, α-chain and β-chain. It is relevant to point out here that the Hb subunits were isolated by a procedure which does not remove the heme group from the polypeptide chain. Furthermore, each subunit was exhaustively purified to ensure the absence of contaminants. In addition, the thiol groups of the α- and β-subunits were completely deprotected following isolation and purification. Therefore, the subunits employed in our investigations do not have any modified amino acid side chains and the data thus obtained should represent an accurate profile of the genetic control of the immune response to Hb and its subunits.

Challenging of Hb-primed mice with either Hb, α- or β-chain demonstrated the existence of gene control in the I-A subregion (Hb and α-chain) and D-end genes. Additional experiments are underway to more precisely map the relevant β-chain specific Ir genes. We have observed that the replacement of a D^s allele with D^r or D^d alleles (and to a less extent D^b alleles) permits responsiveness to the β-subunit in B10.RIII(20R), B10.S(7R) and B10.S(26R) primed using Hb. The data currently available does not allow a statement concerning the role, if any, of Tla or

Qa-region genes in conferring responsiveness to β-chain determinants.

We have postulated that an effective immune response to Hb depends upon the interaction of both I-region and D-end genes. It is conceivable that Ir genes to α-chain and β-chain determinants may interact and effect a modulation of the net immune response to Hb itself. For example, the immune response could be coded by Ir genes in the I-region while genes in the D-end would regulate the response. As a result of such interaction a high responder strain might be converted into a low responder and vice versa. For example, when the I region contains alleles from the $H\text{-}2^q$ haplotype, regardless of the D-end genes, these strains remain low responders suggesting that $H\text{-}2^q$ expresses a true low responder allele at the I-region. However, should the I-region alleles be derived from the $H\text{-}2^s$ haplotype, the insertion of different D alleles can convert these strains into high responders. Therefore, the $H\text{-}2^s$ haplotype actually codes for high responder alleles in the I-subregion but the expression of D^s alleles mediates unresponsiveness. The replacement of D^s allele with D^d or D^r alleles permits an effective immune response. In effect, this suggests that antigenic sites on the α-chain subunit may elicit non-responsive signals to the immune system. A cellular basis for this assertion is being sought.

On the other hand, the immune response to Hb might be under the control of two immune response genes, one mapping in the I region and one mapping in the D-end. The net immune response measured would depend upon complementation between these two loci. This complementation would be similar to the coupled complementation postulated by Dorf and Benacerraf (30). Thus certain interactions would elicit high responses while other interactions would result in decreased responses. A similar Ir/D-end gene interaction has been reported in the autoimmune response to thyroglobulin.

The most interesting observation obtained using Hb as an immunogen was the finding that in some instances the inability to respond to the β-subunit might be attributable to a negative regulatory effect mediated by the α-subunit. Intrepretation of the data was difficult since mice had been primed to both subunits simultaneously. Therefore, mice were primed and challenged to each subunit independently.

The I-A subregion control of immune responsiveness was confirmed by the low responsiveness of B10.A(4R) and B10.MBR and the high responsiveness of B10. However, an alternative explanation for the data is two gene control. It is possible that A^bE^k can interact in B10.A(3R) and B10.A(5R) to generate an immune response but in the reciprocal combination of A^kE^b as tested in B10.A(4R) the interaction is ineffective and no detectable response is generated. We are currently addressing this question using a battery of recombinant F_1 mice in tests for cis- and trans-complementation.

The pattern of responder/low responder phenotypes to α-chain challenge was the same using Hb or α-chain as the immunogen. There was no conversion of α-chain low responder to α-chain responders following priming with α-chain in the absence of the β-subunit. This suggests that the β-subunit does not exert a negative regulatory effect upon the immune responses to the α-chain.

The high responsiveness of B10.S(9R) and the low responsiveness of B10.TL maps β-chain specific Ir genes to the I-A subregion. However, the possibility of the existence of dual gene control has not been ruled out. The pattern of responsiveness to challenge using β-chain was essentially the same observed using Hb primed mice. However, the notable exception was mice of the $H\text{-}2^s$ haplotype. B10.S mice responded strongly to challenge after β-chain priming but only weakly to β-chain subsequent to Hb priming. This observation is consistent with the hypothesis that α-chain determinants can negatively influence the immune response to β-chain in $H\text{-}2^s$ haplotype mice.

The identification of an I-region mutant ($B6.C\text{-}H\text{-}2^{bm12}$) has made investigations into the molecular basis of Ia molecular function. The bm12 mutation expressed in $B6.C\text{-}H\text{-}2^{bm12}$ mice is thought to be expressed in the A_β subunit of A^b molecules such that the Ia.8 specificity is lost (31). Under conditions in which parental B6/Kh ($H\text{-}2^b$) mice respond to challenge using α-chain of Hb, bm12 mutant mice cannot. Thus, the loss of Ia.8 specificity alters immune reactivity towards the α-subunit. Whether Ia.8 is necessary for antigen binding and presentation to T-cells or whether Ia.8 is necessary for recognition of the α-chain sites as altered self via single or dual receptors is not yet known.

We also have been able to demonstrate there exists non-H-2 genes that influence the immune response to the β-chain of Hb. A second, but somewhat weaker effect, is also observed for in vitro responses to the α-subunit. Balb/c mice are almost non-responsive to the β-subunit of Hb under conditions in which B10.D2 mice respond. When challenged and primed using the non-cross-reacting α-chain, Balb/c mice respond at a 50% level to that observed for B10.D2 mice. Experimentation is currently underway to enumerate and map these non-H-2 genes.

At least one non-H-2 gene that is required for an in vitro response to either subunit is associated with the Ly-1 locus. Depletion of Ly-1$^+$ T-cells but not Ly-2$^+$ T-cells nearly completely inhibited the in vitro T-cell proliferative responses. We have not been successful in converting low responder mice to responders by depletion of Ly-2$^+$ cells.

In future experiments we plan to utilize banks of T-cell clones and monoclonal anti-Hb antibodies (currently available in our laboratories to identify the immunologically relevant sites on the Hb molecule and to construct a detailed network of cell-cell communications involved in the immune response to this molecule.

ACKNOWLEDGEMENTS

This work was supported by grants AM-18920 from the National Institute of Arthritis and Metabolic Diseases, AI-18657 from the National Institute of Allergy and Infectious Diseases and CA-24473 from the National Cancer Institute, National Institutes of Health, U.S. Public Health Service.

REFERENCES

1. C.J. Krco and C.S. David. Genetics of immune responses: A survey. CRC Critical Rev. Immunol. 1,211 (1981).
2. E.C. Lozner, D.H. Sachs, and G.M. Shearer. Genetic control of the immune response to staphylococcal nuclease. I. Ir-nase control of the antibody response to nuclease by the Ir region of the mouse H-2 complex. J. Exp. Med. 139,1204 (1974).
3. K. Okuda, P. Christadoss, S. Twining, M.Z. Atassi, and C.S. David. Genetic control of immune response to sperm whale myoglobin in mice. I. T-lymphocyte proliferative response under H-2 linked Ir gene control. J. Immunol. 121,866 (1978).
4. K. Okuda, S.S. Twining, C.S. David, and M.Z. Atassi. Genetic control of immune response to sperm whale myoglobin in mice. II. T-lymphocyte proliferative response to the synthetic antigenic sites. J. Immunol. 23,182 (1979).
5. M.Z. Berzofsky. Genetic control of the immune response to mammalian myoglobins in mice. I. More than one I-region in H-2 controls the antibody response. J. Immunol. 120,260 (1978)
6. S. Yokota, C.S. David and M.Z. Atassi. Genetic control of the immune response to myoglobin. III. Autoimmune T-lymphocyte proliferative response to mouse myoglobin. Mol. Immunol. 17,1079 (1980).
7. S.S. Twining, C.S. David and M.Z. Atassi. Genetic control of the immune response to myoglobin. IV. Mouse antibodies in outbred and congenic strains against sperm-whale myoglobin recognize the same antigenic sites that are recognized by antibodies raised in other species. Mol. Immunol. 18,447 (1981).

8. M.Z. Atassi, S.S. Twining, H. Lehmann and C.S. David. Genetic control of the immune response to myoglobin. V. Analysis of the cross-reactivity of 12 myoglobins with sperm-whale myoglobin antisera of inbred mouse strains in terms of substitutions in the antigenic sites and in the environmental residues of the sites. Immunol. Commun. 10,359 (1981).
9. M.Z. Atassi, S. Yokota, S.S. Twining, H. Lehman and C.S. David Genetic control of the immune response to myoglobin. VI. Inter-site influences in T-lymphocyte proliferative response from analysis of cross-reactions of ten myoglobins in terms of substitutions in the antigenic sites and in environmental residues of the sites. Mol. Immunol. 18,945 (1981).
10. R.H. Schwartz, A.M. Solinger, M. Utlee and E. Margoliash. In Immunobiology of Proteins and Peptides (edited by M.Z. Atassi and A.B. Stavitsky) p. 371, Plenum Press, New York (1978).
11. E.E. Sercarz, R.L. Yowel, D. Turkin, A. Miller, B.A. Araneo and L. Adorini. Different functional specificity repertories for suppressor and helper T-cells. Immunological Rev. 39,108 (1978).
12. K. Okuda, S. Sakata, M.Z. Atassi and C.S. David. Genetic control of the immune response to hen's egg-white lysozyme in mice. I. Antibody and T-lymphocyte proliferative responses to the native protein. J. Immunogen. 6,447 (1979).
13. I. Melchers, and K. Rajewsky. Specific control of responsiveness by two complementing Ir loci in the H-2 complex. Eur. J. Immunol. 5,753 (1975).
14. I. Melchers, K. Rajewsky and D.C. Shreffler. Ir-LDH$_B$: map position and functional analysis. Eur. J. Immunol. 3,754. (1973).
15. L.J. Rosenwasser, M.Z. Barcinski, R.H. Schwartz, and A.S. Rosenthal. Immune response gene control of determinant selection. II. Genetic control of the murine T lymphocyte proliferative response to insulin. J. Immunol. 123,471 (1979).
16. P. Christadoss, Y-C.M. Kong, M. Elrehewy, N.R. Rose and C.S. David. Genetic control of T-lymphocyte proliferative autoimmune response to thyroglobulin in mice. In: Genetic Control of Autoimmune Disease (edited by Rose, Bigazzi and Warner) p. 445, Elsevier-North Holland. (1978).
17. Y-C.M. Kong, C.S. David, A.A. Giraldo, M. Elrehewy and N.R. Rose. Regulation of autoimmune response to mouse thyroglobulin: influence of H-2D-end genes. J. Immunol. 123,5 (1979).
18. A.L. Kazim and M.Z. Atassi. Prediction and confirmation by synthesis of two antigenic sites in human hemoglobin by extrapolation from the known antigenic structure of sperm-whale myoglobin. Biochem. J. 167,275 (1977).
19. A.L. Kazim and M.Z. Atassi. A novel and comprehensive synthetic approach for the elucidation of protein antigenic structures: Determination of the full antigenic profile of the α-chain of human hemoglobin. Biochem. J. 191,261 (1980).

20. C.J. Krco, A.L. Kazim, M.Z. Atassi and C.S. David. Genetic control of the immune response to hemoglobin. I. Demonstration of separate genetic control of the responses to the α- and β-subunits by in vitro lymphocyte proliferation. J. Immunogenet. 8,315 (1981).
21. C.J. Krco, A.L. Kazim, M.Z. Atassi and C.S. David. Genetic control of the immune response to hemoglobin. II. Studies using purified α-chain and β-chain as immunogens. J. Immunogenet. 8,395 (1981).
22. C.J. Krco, A.L. Kazim, M.Z. Atassi, R. Melvold and C.S. David Genetic control of the immune response to hemoglobin. III. Variant A_β (bm12) but not A_e (D2.GD) Ia polypeptides alter immune responsiveness towards the α-subunit of human hemoglobin. J. Immunogenet. (in press) (1981).
23. C.R. Young. Genetic control of the immune response to ferritin in mice. J. Immunogenet. 3,199 (1976).
24. C.R. Young. Genetic control of the immune response to mammalian chymotrypsins in mice. J. Immunogenet. 6,45 (1979).
25. N.M. Vaz, E.M. Vaz and B.B. Levin. Relationship between histocompatibility (H-2) and genotype and immune responsiveness to low doses of ovalbumin in the mouse. J. Immunol. 104,1572 (1970).
26. J.A. Marsh, T.E. Wheat and E. Goldberg. Temporal regulation of the immune response to LDH-C_4 by an X-linked gene in C3H/HeJ and SJL/J mice. J. Immunol. 118,2293 (1977).
27. H. Nowack, E. Hahn, C.S. David, R. Timpl and D. Gotze. Immune response to calf collagen Type I in mice: a combined control of Ir-IA and non-H-2 linked genes. Immunogenet. 2,331 (1975).
28. D.H. Sachs, J.Z. Berzofsky, D.C. Pisetsky and R.H. Schwartz Genetic control of the immune response to staphylococcal nuclease. Springer Semin. Immunopath. 1,51 (1978).
29. C.J. Krco, A.L. Kazim, M.Z. Atassi and C.S. David. Genetic control of the immune response to hemoglobin. IV. Ly-1^+ T-cells and appropriate non-H-2 genes are required for in vitro responses to α- and β-subunits of human adult hemoglobin. J. Immunogenet. 8,471 (1981).
30. M.E. Dorf and B. Benacerraf. Complementation of H-2 linked Ir genes in the mouse. Proc. Natl. Acad. Sci. 72,3671 (1975).
31. W.P. Lafuse, J.F. McCormick, R.W. Melvold, and C.S. David. Serological and biochemical analysis of Ia molecules in the I-A mutant B6.C-H-2^{bm12}. Transpl. 31,434 (1981).

8

IMMUNE RESPONSE GENE CONTROL OF T DEPENDENT REACTIVITY TO COLLAGEN IN MAN

Alan M. Solinger and John D. Stobo

Section of Rheumatology/Clinical Immunology
Dept. of Medicine and The Howard Hughes Medical Institute
University of California, San Francisco
San Francisco, CA 94143

INTRODUCTION

T dependent reactivity controlled by immune response (Ir) genes in rodents exhibits the following three characteristics. (1-3) First, it is dependent on interactions between T cells and antigen presenting accessory cells (conveniently referred to here as MØ). Second, the reactivity is directed against determinants inherent in either synthetic polypeptides or limited regions of globular proteins. Third, the reactivity is controlled by genes present in the immune response (Ir) locus of the major histocompatibility complex.

While it has been assumed that Ir genes also exist in man, much of the evidence supporting this has been indirect and derives from studies demonstrating that products of genes in the HLA-D region of the human major histocompatibility complex demonstrate a tissue localization, ability to stimulate in a mixed lymphocyte reaction, and an amino acid sequence similar to that documented for rodent Ir gene products (reviewed in 4) Recently, direct evidence has been presented suggesting an association between HLA-D linked genes and the ability to respond or not respond to certain antigens. (5,6)

In this present study we investigate the documented association between HLA-DR4, rheumatoid arthritis and T dependent reactivity to collagen to determine if genes linked to those coding for HLA-DR4 represent collagen Ir genes. (7-11) The studies indicate that the cellular, molecular, and genetic characteristics established for T dependent reactivity controlled by Ir genes in rodents parallels that observed for collagen reactivity in man.

That is to say, the reactivity requires interactions between MØ and T cells and could be elicited by determinants displayed by collagen-like peptides lacking secondary or tertiary structure. Most importantly, all HLA-DR4 positive individuals, either normals or patients with rheumatoid arthritis, are collagen responders. In addition, preliminary studies suggest that absence of discernible collagen reactivity in HLA-DR^{-} non-responders reflects a predominance of collagen specific, radiation-sensitive, T-dependent suppressive influences.

MATERIALS AND METHODS

The patient population consisted of 20 individuals with definite rheumatoid arthritis, as defined by criteria of the American Rheumatism Association. Patients receiving corticosteroids, cytotoxic drugs, or penicillamine were not included. Normals consisted of volunteers chosen at random from a pool of blood donors.

Antigen reactivity was determined utilizing an indirect assay for the production of leukocyte inhibition factor (LIF) as previously described. (12) Peripheral blood mononuclear cells (PBMC) obtained by sedimenting peripheral blood over ficoll-hypaque were incubated with either antigen or media (RPMI-1640) for two hours at 37 C on a rocker platform. The cells were then washed three times with RPMI-1640, resuspended in 10% fetal calf serum, and incubated without any addition of antigen at 37 C for 48 hours. The culture fluids were then removed and assayed for their ability to inhibit random migration of normal neutrophils during a 24 hour incubation. The results are presented as % inhibition of migration calculated by the formula

$$\left[1 - \left(\frac{\text{area of migration in culture fluids from cells initially incubated with antigen}}{\text{area of migration in culture fluids from cells initially incubated without antigen}}\right)\right] \times 100$$

Twenty percent inhibition of migration was chosen as delineating antigen responsiveness from antigen unresponsiveness. This was based on our observation that individuals demonstrating positive delayed hypersensitivity skin tests to Candida albicans synthesized sufficient LIF in response to in vitro challenge with this antigen to inhibit neutrophil migration by 20%. In contrast, individuals who failed to respond to delayed hypersensitivity skin testing with Candida albicans synthesized amounts of LIF which caused less than 20% inhibition of migration. It should be noted that direct addition of antigen to neutrophils failed to affect their migration, arguing against antigen carryover as a consideration in the results. Additionally, four concentrations of antigen were used to ensure that any observed absence of reactivity did not simply reflect the use of a suboptimal concentration. The antigens used to elicit LIF production included commercial beef collagen (Sigma Chemical Co., St. Louis, Missouri) heat denatured as previously described (13), Candida albicans (C. albicans, Hollister-Steir Labs, Toronto, Ontario dialyzed against phosphate buffered saline, pH 7.4), the synthetic polypeptides $(Gly\text{-}Pro)_n$ and $(Pro)_n$ (generous gifts of Dr. Rajendra Bhatnagar), chick type I, type II collagen prepared from lathrytic chick embryos, and proteoglycan. (These latter three reagents were generous gifts of Dr. Robert Siegel). HLA-DR typing was performed in the laboratory of Dr. Paul I. Terasaki utilizing a panel of 38 reagents which depict 6 HLADR determinants (HLA-DR 1-5 and 7).

The specificities of two monoclonal antibodies used to characterize mononuclear cells required for antigen induced LIF production have been previously described. (14,15) Briefly, T-29 is a κ, γ_2b immunoglobulin with specificity for 10 ± 3% of the peripheral blood T cells from several HLA-DR disparate individuals. T-29 does not react with B cells, null cells, or macrophages (MØ). Mac-120 is a κ, μ immunoglobulin with specificity for 38 ± 4% of the adherent, peripheral blood MØ from several HLA-DR disparate subjects. Mac-120 reacts only with MØ.

RESULTS

PBMC from a total of 20 patients with rheumatoid arthritis and 54 normal individuals were assayed for their ability to synthesize the lymphokine, LIF, in response to challenge with denatured beef collagen or C. albicans. All individuals were responders to C. albicans as defined by the synthesis of LIF sufficient to cause greater than 20% inhibition of leukocyte migration. Utilizing similar criteria, 18/20 patients with rheumatoid arthritis and 18/54 normal individuals could be defined as collagen responders. The reactivity of the total 36 collagen responders and 38 collagen non-responders to both collagen and C. albicans is summarized in Table 1. It should be emphasized that the collagen non-responders failed to respond to any of four concentrations of collagen used. This argues against the possibility that the absence of reactivity reflected the use of an inappropriate concentration of antigen.

To determine the nature of the cells required for the depicted collagen reactivity we used two monoclonal antibodies (T-29, Mac-120) which we have previously demonstrated to be reactive with mononuclear cells required for T proliferative responses to other conventional antigens (C. albicans, purified protein derivative; 14,15) T-29 is capable of cytolytically removing those T cells required for antigen induced T cell proliferation while Mac-120

TABLE 1

ANTIGEN INDUCED LIF PRODUCTION IN COLLAGEN

RESPONDERS AND NON-RESPONDERS

Collagen Responder Status	% Inhibition of Migration in Response to	
	Denatured Beef Collagen	C. Albicans
Responder (36)[2]	38 ± 13	41 ± 12
Non-Responder (38)	34 ± 12	5 ± 12

[1]Results presented as arithmetic mean inhibition of leukocyte migration ± S.E.

[2]Numbers in parentheses = individuals studied

can cytolytically remove a subpopulation of HLA-DR bearing MØ involved in the presentation of soluble antigen to reactive T cells. As demonstrated in Table 2, PBMC from two normal collagen responders remaining after cytolysis with either T-29 or Mac-120 were unresponsive to both C. albicans and bovine collagen. Addition of T-29 treated PBMC to Mac-120 treated PBMC resulted in restoration of the reactivity to collagen (Exp #2). These experiments indicate that the cellular requirements for collagen induced LIF production are the same as those required for T dependent reactivity to other conventional antigens.

To investigate the nature of the antigenic determinants in collagen recognized by the reactive T cells, the experiments outlined in Table 3 were performed. PBMC from 10 normal individuals responsive to denatured bovine collagen were tested for their reactivity to chick type I and type II collagen purified to homogeneity as determined by SDS-PAGE electrophoresis, the synthetic polypeptide $(Gly\text{-}Pro)_n$ (mean molecular weight of 11,000 daltons) which lacks collagen like helical structure as determined by optical rotatory dispersion and circular dichromism, the synthetic polypeptide $(Pro)_n$ (mean molecular weight of 37,000 daltons) and proteoglycan, purified to homogenicity as determined by SDS-PAGE electrophoresis. Chick type I and type II collagen elicited reactivity equivalent to that seen with the denatured beef collagen. This suggests that reactivity was directed against a basic immunodominant determinant common to different types of collagen. That this determinant may be inherent in linear sequences of (Gly-Pro) is demonstrated by the finding that the synthetic polypeptide $(Gly\text{-}Pro)_n$ was as effective as the denatured beef collagen in inducing LIF production. Neither $(Pro)_n$ nor proteoglycan elicited reactivity among collagen responders. $(Pro)_n$ did not inhibit LIF production occurring in response to C. albicans ruling against any toxic effect of the polypeptide. None of the collagen preparations or polypeptides elicited reactivity among collagen non-responders.

To this point, the data demonstrate that the cellular and molecular characteristics of reactivity to collagen in man are similar to those demonstrated for T dependent reactivities controlled by Ir genes in rodents. That is to say, collagen induced synthesis of LIF requires interactions between antigen presenting MØ and T cells and occurs in response to antigenic determinants inherent in a linear polypeptide. To determine if this collagen reactivity is associated with genes linked to those in the proposed human Ir locus (i.e. HLA-D), PBMC from 53 individuals were simultaneously tested for reactivity to denatured bovine collagen and typed for 7 HLA-DR determinants (Table 4). Of the 28 collagen responders, 21 were HLA-DR4 positive. This included 11 of 15 normal collagen responders. Of the 21 HLA-DR4 positive individuals, all were collagen responders. Statistical analysis of this

TABLE 2

REQUIREMENT FOR T CELLS AND MØ IN THE PRODUCTION OF LIF BY C. ALBICANS AND BOVINE COLLAGEN

Treatment of Responder Cells	% Inhibition of Migration in Response to	
	C. albicans	Bovine Collagen
Exp #1		
Control[1]	81	43
Anti-T[2]	0	0
Anti-MØ[3]	15	11
Exp #2		
Control	55	49
Anti-T	0	2
Anti-MØ	19	14
Anti-T + Anti-MØ[4]	N.D.	42

[1]Control = cytolysis with either MOPC 195 (κ, γ_2b) or TEPC 183 (κ, μ)

[2]Anti-T = cytolysis with the κ, γ_2b monoclonal antibody T-29

[3]Anti-MØ = cytolysis with the κ, μ monoclonal antibody Mac-120

[4]Refers to populations treated either with T-29 or Mac-120 and then combined

N.D. = not done

TABLE 3

REACTIVITY OF COLLAGEN RESPONDERS AND NON-RESPONDERS TO COLLAGEN AND RELATED ANTIGENS

	% Inhibition of Migration in Response To:					
	Bovine Collagen	Chick[1] Type I	Chick[2] Type II	$(Gly\text{-}Pro)_n$	$(Pro)_n$	Proteoglycan
Responders (10)	41±4.2	45±4.6	38±6.7	39±3.2	6.3±3.9	9.0±4.1
Non-Responders (3)	11±3.6	6.4±2.8	12.1±3.1	9.0±4.0	6.9±1.1	3.3±2.9

[1] Purified chick type I collagen

[2] Purified chick type II collagen

TABLE 4

ASSOCIATION BETWEEN COLLAGEN REACTIVITY AND HLA-DR4 POSITIVITY IN PATIENTS WITH RHEUMATOID ARTHRITIS AND NORMALS

	Reactivity to Collagen			
	Responder[1]		Non-Responder	
HLA-DR4 Positivity	R.A.[2]	Normal	R.A.	Normal
+	10	11	0	0
-	3	4	2	23

[1]Collagen induced production of LIF sufficient to cause greater than 20% inhibition of migration

[2]Rheumatoid arthritis

association between HLA-DR4 positivity and collagen reactivity demonstrated a P value of less than 0.0001 and a chi square of 31.1.

These studies suggest that genes linked to those coding for HLA-DR4 constitute human Ir genes for reactivity to (Gly-Pro) determinants inherent in collagen. However, this conclusion is valid only if the absence of discernible reactivity noted in collagen non-responders represents the absence of collagen reactive T cells. In other words, it is possible that in HLA-DR4 negative individuals, reactive T cells co-exist with specific suppressive influences. The lack of apparent reactivity might represent the preponderance of these suppressive influences rather than any absence of collagen reactive T cells. To test this possibility, we took advantage of the fact that some suppressive influences are radiosensitive, while LIF production which requires only protein synthesis is relatively radio-resistant. Peripheral blood mononuclear cells from 5 HLA-DR4$^-$, collagen non-responders were irradiated with 1000 rads and compared with non-irradiated cells for their LIF production in response to collagen and C. albicans (Table 5). In each case irradiation resulted in the appearance of collagen reactivity. That this did not represent a non-specific effect of irradiation is suggested by the finding that irradiation did not result in a substantial increase in reactivity to C. albicans. Moreover, in data not presented it could be demonstrated that irradiation did not result in the acquisition of responder status in two individuals unresponsive to PPD.

If the appearance of collagen responsiveness among the irradiated cells indeed represented inactivation of collagen specific suppressive influences, then addition of the unresponsive non-irradiated cells to autologous, irradiated reactive cells should specifically inhibit LIF production occurring in response to collagen. The data from preliminary experiments to determine this were performed in the following way (Table 6). Non-irradiated cells from a HLA-DR4$^-$ collagen non-responder were fractionated into B enriched (80% Ig bearing, 10% T) and T enriched (96% E rosette forming, < 1% MØ) populations by a combination of adherence and fractionation of E rosette forming cells. An increasing number of either the B or T population was then added to a decreasing number of irradiated, autologous, collagen responsive cells so that the final number of cells in each culture was comparable. The reactivity of these cultures to collagen and C. albicans was then compared (Table 6). In this situation, the B enriched population is unresponsive to antigen because it lacks reactive T cells, while the T enriched population is unresponsive because it was sufficiently depleted of antigen presenting cells. Thus, the B population should serve as a control to indicate "dilution" of reactivity. At each ratio tested, collagen reactivity among

TABLE 5

EFFECT OF IRRADIATION ON REACTIVITY TO COLLAGEN AND C. ALBICANS IN HLA-DR4$^-$ COLLAGEN NON-RESPONDERS

		% Inhibition of Migration in Response to	
		Collagen	C. albicans
#1	Pre-irrad.	3	36
	Post-irrad.[1]	48	32
#2	Pre-irrad.	13	52
	Post-irrad.	71	63
#3	Pre-irrad.	0	41
	Post-irrad.	49	60
#4	Pre-irrad.	8	58
	Post-irrad.	63	43
#5	Pre-irrad.	11	62
	Post-irrad.	36	54

[1]After in vitro irradiation with 1000 rads.

TABLE 6

EFFECT OF NON-IRRADIATED B CELLS OR T CELLS ON COLLAGEN AND C. ALBICANS REACTIVITY AMONG IRRADIATED AUTOLOGOUS PBMC

Cell Mixtures Assayed	% Inhibition of Migration in Response to	
	Collagen	C. albicans
Irrad. PBMC[1] + Non-Irrad.		
4×10^6 + 0	52	43
3×10^6 + 1×10^6 B	44	52
2×10^6 + 2×10^6 B	35	32
1×10^6 + 1×10^6 B	28	18
0 + 4×10^6 B	3	2
3×10^6 + 1×10^6 T	22	41
2×10^6 + 2×10^6 T	10	40
1×10^6 + 3×10^6 T	0	45
0 + 4×10^6 T	0	0

[1]Irradiated with 1000 rad.

populations containing PBMC and non-irradiated T cells was less than that noted for PBMC plus non-irradiated B cells. In other words, when compared to the effects of dilution mediated by the intrinsically unresponsive B population, non-irradiated, unresponsive T cells suppressed collagen reactivity. This suppression was not noted when the reactivity of the cell mixtures to C. albicans was tested. The unresponsive, non-irradiated, T enriched, MØ depleted population did not "dilute" reactivity since MØ in the irradiated PBMC could serve to present antigen to the irradiated T cells. (We have studied one individual who was initially unresponsive to C. albicans and in whom irradiation of the PBMC resulted in the appearance of reactivity. In this individual, non-irradiated T cells did suppress reactivity among the irradiated PBMC).

DISCUSSION

The data presented here indicate that T dependent reactivity to collagen in man is under Ir gene control. This conclusion is supported by the following three findings. First, LIF production induced by collagen requires interactions between antigen presenting accessory cells and antigen responsive T cells which are also required for reactivity to other, conventional antigens. Second, reactivity is directed against determinants inherent in denatured collagen and which are expressed by the linear, synthetic polypeptide (Gly-Pro). Induction of LIF synthesis is not dependent on conformational determinants. Third, in both normal individuals as well as patients with rheumatoid arthritis reactivity to collagen is significantly associated with genes linked to those in the proposed, human Ir locus. In other words, the cellular, molecular, and genetic characteristics found for collagen induced production of LIF in man are similar to those established for T dependent reactivities controlled by Ir genes in rodents.

Our findings should be discussed within the framework provided by the studies performed in rodents and the recently reported studies of Sasazuki et al in man. The majority of individuals we assayed for collagen reactivity are heterozygous at the HLA-D locus. Our initial finding of collagen reactivity in all HLA-DR4^{+} heterozygotes demonstrates that collagen responsiveness is dominant. (One individual who was homozygous for HLA-DR4 demonstrated reactivity to collagen which was not significantly higher than that seen in heterozygotes). This is certainly consistent with the dominance of Ir genes demonstrated in rodents. However, the data presented in Tables 5 and 6 suggest that all individuals are collagen responders and that our inability to detect collagen induced LIF production among HLA-DR4^{-} individuals reflects the presence of radio-sensitive, T-dependent suppressive influences. In data not presented it could be demonstrated that radio-sensitive influences capable of suppressing collagen reactivity are associated with all HLA-DR types studied except HLA-DR4 (i.e. HLA-DR 1, 2, 3, 5, 6, and 8). These suppressive influences are not dominant in that heterozygous individuals whose MHC contains genes coding for HLA-DR4 as well as genes coding for another HLA-DR type associated with suppressive influences are collagen responders. This is quite different than the simple model of dominant inheritance noted for immune suppressor genes documented in some rodent models and the dominance of immune suppressor genes delineated by Sasazuki et al in man. (5,16)

Our data is compatible with the thesis that either during the physiologic metabolism of collagen or during ingestion of certain foods all individuals are exposed to (Gly-Pro) determinants normally sequestered in the collagen triple helix. In all individuals except those whose MHC contains genes linked to those coding for HLA-DR4 this results in the activation of both collagen reactive and collagen suppressive cells with the latter predominating. In HLA-DR4$^+$ individuals, only collagen reactive and not collagen suppressive cells are generated. (This latter conclusion is based on data demonstrating that irradiation of PBMC from HLA-DR4$^+$ individuals did not result in any increase in collagen reactivity. (A. Solinger and J. Stobo, submitted for publication). Whether this represents a situation in which HLA-DR4 linked genes code for events which actively inhibit either the reactivity or the generation of suppressive influences or whether it indicates that two genes are required to act in a trans-complementary fashion to generate suppressive influences is not clear. In this latter situation the complementing genes would be linked to all HLA-D genes except those coding for HLA-DR4.

Our studies do not allow any definite conclusions concerning the relationship between reactivity to collagen and the development of rheumatoid arthritis. They do, however, suggest that while reactivity to collagen may be necessary for the development of rheumatoid arthritis it is clearly not by itself sufficient.

ACKNOWLEDGEMENTS

We thank Dr. R. Bhatnagar for his gifts of collagen related peptides and Dr. R. Siegel for the purified chick collagen and proteoglycan.

REFERENCES

1. A. Rosenthal, M. Barginski, and L. Rosenwasser. Immune response gene control of determinant selection. J. Immunol. 123:471 (1979)
2. W. Paul and B. Benacerraf. Functional specificity of thymus dependent lymphocytes. Science 195:1293 (1977)
3. H. McDevitt and A. Chinitz. Genetic control of antibody response: Relationship between immune response and histocompatability (H-2) type. Science 163:1207 (1969)
4. R. Winchester and H. Kunkel. The human Ia system. In: Advances in Immunology, F. Dixon and H. Kunkel eds., Academic Press, New York, 28: 222 (1979)
5. T. Sasazuki, N. Ohta, R. Kaneoka, and S. Kojima. Association between an HLA haplotype and low responsiveness to schistosomal worm antigen in man. J. Exp. Med. 152:3145 (1980)
6. W. Griffin, S. Moore, H. Luthra, C. McKenna, and C. Fathman. Association of antibodies to native DNA with HLA-DRw3. A possible major histocompatability complex-linked human immune response gene. J. Exp. Med. 152: 3195 (1980)
7. D. Trentham, R. Dynesius, R. Rocklin, and J. David. Cellular sensitivity to collagen in rheumatoid arthritis. New Eng. J. Med. 299:327 (1978)
8. J. Stuart, A. Postlethwaite, A. Townes, and A. Kang. Cell mediated immunity to collagen and collagen chains in rheumatoid arthritis and other rheumatic diseases. Amer. J. Med. 69:13 (1980)
9. J. Smolen, E. Menzel, O. Scherak, M. Kojer, G. Kolarz, C. Steffen, and W. Mayr. Lymphocyte transformation to denatured type I collagen and B lymphocyte alloantigens in rheumatoid arthritis. Arth. and Rheum. 23:424 (1980)
10. A. Gibofsky, R. Winchester, E. Yunis, and H. Kunkel. Contrasting patterns of newer histocompatibility determinants in patients with rheumatoid arthritis and systemic lupus erythematosus. Arth. and Rheum. 21: 1345 (1978)
11. P. Stastny. HLA-D and Ia antigens in rheumatoid arthritis and systemic lupus erythematosus. Arth. and Rheum. 21:1395 (1978)
12. R. Rocklin. Products of activated lymphocytes: Leukocyte inhibitory factor (LIF) distinct from migration inhibitory factor (MIF). J. Immunol. 112:1461 (1974)
13. L. Rosenwasser, R. Bhatnagar, and J. Stobo. Genetic control of the murine T lymphocyte proliferative response to collagen: Analysis of the molecular and cellular contributions to immunogenicity. J. Immunol. 124:2854 (1980)
14. P. Hausman, H. Raff, R. Gilbert, L. Picker, and J. Stobo. T cells and macrophages involved in the autologous mixed lymphocyte reaction are required for the response to conventional antigen. J. Immunol. 125:1374 (1980)

15. H. Raff, L. Picker, and J. Stobo. Macrophage heterogeneity in man. J. Exp. Med. 152:581 (1980)
16. B. Benacerraf and R. Germain. Specific suppressor responses to antigen under I region control. Fed. Proc. 38:2053 (1979)

REGULATION OF THE IMMUNE RESPONSE

9

MYOGLOBIN-REACTIVE T CELL CLONES

A.J. Infante*, M.Z. Atassi+ and C.G. Fathman*

*Division of Immunology, Stanford University Medical Center, Stanford, California, 94305.
+Division of Immunology, Mayo Clinic, Rochester, Minnesota, 55901.

Myoglobin has been used for a considerable period of time as a "model" antigen for immunological studies. Since the sequence and crystal structure of myoglobin from sperm whale skeletal muscle are well known, myoglobin can be studied in great detail as an antigen. The immunological studies that have been done with Mb to date have provided useful information bearing on a number of important questions. We have used myoglobin as a model antigen to study via the new and powerful techniques made possible by T cell cloning. Since the T cell clones which we utilize are of the helper/inducer subset (T_H, see below) and therefore occupy a central role *in vivo* in controlling immune responsiveness, these techniques should provide both specific and generalized information relating to the immune responses to complex protein antigens.

Myoglobin is a "multideterminant" antigen. Serum antibodies have been shown to recognize five discrete sites consisting of six or seven amino acids each (1). Myoglobin antibody-binding sites are continuous stretches of the polypeptide chain but the native protein conformation is necessary for maximal antibody binding. There are also multiple sites on myoglobin which are recognized by T-cells (2), but their locations are not clearly determined, being defined by cyanogen bromide (CNBr) fragments. There are three CNBr fragments making up the entire Mb molecule: residues 1-55, 56-131 and 132-153.

The immune response to Mb is controlled in the mouse by Ir genes mapping within the major histocompatibility complex (MHC), H-2 (3). These responses have been mapped using congenic and recombinant mouse strains to study *in vivo* antibody formation and

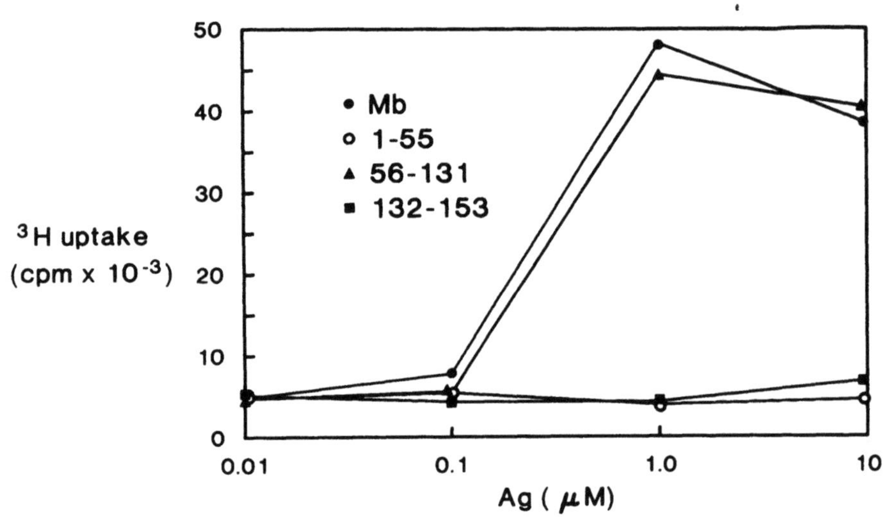

FIGURE 1

CLONE 26.17 RECOGNIZES A DETERMINANT IN FRAGMENT 56-131
Clone 26.17 was cultured with varying concentrations of Mb or its CNBr fragments. Proliferation was measured on day 3.

in vitro T cell proliferation. It has been found that both antibody and T-cell responses map to both the I-A and I-E/C subregions of H-2. It has been presumed that reponses to different Mb determinants are controlled by particular Ir genes mapping to either I-A or I-E/C.

Techniques developed in our laboratory have permitted the isolation and analysis of T_H cell clones reactive to soluble synthetic and naturally occurring polypeptide antigens (4-6). These T_H cells proliferate in response to specific antigen, but only when recognition of antigen is coupled to recognition of specific Ia antigens on antigen presenting cells (APC). This can be shown by Ia "restriction" of recognition on various congenic and recombinant strain APC (restriction mapping) and blocking of T cell reactions by anti-Ia monoclonal antibodies (antibody blocking).

We now present a series of myoglobin-reactive T_H cell clones (5) and a commentary based on the results of analysis of their proliferation and Ia restriction. These clones were isolated from Mb-primed lymph node cells of F_1 mice of strains C57BL/6 (B6) and A/J(A). The F_1 mice are designated (B6A)F_1. Lymph node cells were propagated by serial restimulation with antigen and syngeneic irradiated spleen cells as a source of APC. These long-term cell lines were then cloned by a combination of soft agar colony formation and liquid culture-limiting dilution methods. Proliferation was measured in triplicate cultures of cloned T cells with antigen and irradiated spleen cells as a source of APC. Cultures were pulsed with ^{3}HTdR and counted after collection on filter paper. Responses were blocked by adding microgram amounts of purified monoclonal anti-Ia antibodies to cell cultures.

Clone 26.17

Clone 26.17 illustrates several salient points regarding the response of T_H cells to Mb. Clone 26.17 proliferates in response to myoglobin and CNBr fragment 56-131. Proliferation is not induced by fragments 1-55 and 132-153, thus localizing a T_H cell recognition site to residues 56-131. The dose response curves produced by stimulation of clone 26.17 with Mb or 56-131 are identical. This suggests that T_H cells do not recognize antigen in its native conformation, since antibodies (which seem to recognize native Mb) bind 4-10 times more effectively to Mb than to the corresponding CNBr fragments (7). Such cross-reactivity of T_H cells to native and denatured determinants of proteins has been described in other systems (8).

Clone 26.17 is restricted to recognizing Mb presented by (B6A)F_1 APC but not A or B6 strain APC. This clone also reacts to Mb on APC of [B10.A(4R) x A]F_1 mice, mapping the response to

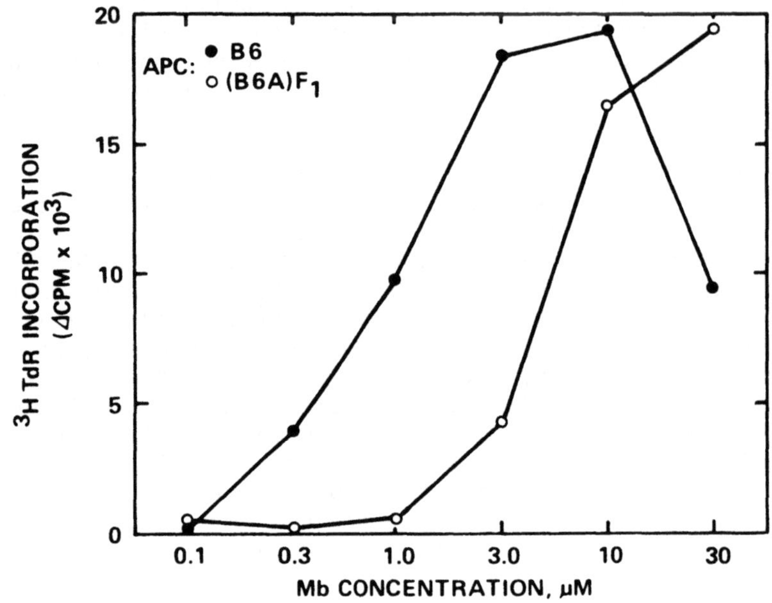

FIGURE 2

Clone 28.1 quantitatively recognizes antigen (myoglobin) and an Ia molecule of the _b_ haplotype. Antigen presenting cells of strain B6 and $(B6A)\overline{F_1}$ were used to present varying concentrations of Mb to clone 28.1

I-A "heterozygous" antigens, either $A^b_\alpha A^k_\beta$ or $A^k_\alpha A^b_\beta$ molecules (monoclonal antibody inhibition suggests restriction to $A^k_\alpha A^b_\beta$ molecules). Thus, clone 26.17 reacts to a specific "combination" of Mb determinant (i.e., 56-131) and Ia antigen ($A^k_\alpha A^b_\beta$). This mapping defines an I region controlled immune response (?Ir gene) to a specific antigen determinant.

Clone 28.1

Clone 28.1 responds to Mb and fragment 1-55 with identical dose response curves. This clone is restricted to $I\text{-}A^b$ determinants (either A^b_α, A^b_β or the complex $A^b_\alpha A^b_\beta$) since it proliferates in response to Mb presented by APC of strains B6, $(B6A)F_1$, and B10.A(3R). These responses (to a determinant in residues 1-55 restricted by $I\text{-}A^b$) define a qualitative immune response. In addition, this clone exhibits a quantitative response to Mb and I-A (9). At every Mb concentration clone 28.1 responds more effectively to Mb presented by strain B6 than to $(B6A)F_1$ APC. Conversely, at a single Mb concentration, clone 28.1 responds more effectively to a given number of strain B6 APC. This suggests that strain B6 APC express more $I\text{-}A^b$ determinants on their cell surface than $(B6A)F_1$ (presumably <u>via</u> a gene dose effect). We confirmed this differential expression of $I\text{-}A^b$ determinants by FACS analysis of B6 and $(B6A)F_1$ cells using an FITC conjugated anti-I-A^b monoclonal antibody.

Clone 45.49

Clone 45.49 responds to fragment 1-55 in a manner identical to clone 28.1, but 45.49 is restricted to Ia antigens of the $A^b_e E^k_\alpha$ type (i.e., I-E region restriction). Restriction mapping and antibody inhibition confirm this result. We have argued that this clone, which cross-reacts to $A^b_e E^d_\alpha$ determinants, explains why Ir gene control of responsiveness to certain Mb determinants was originally mapped to the I-C (10), rather than I-E subregion of H-2. This re-mapping is in keeping with a simplified approach to the genetics of I-region control of immune responses recently proposed by Klein et al (11).

Clone 69

Clone 69 responds to a determinant on 56-131 and is restricted to $A^k_e E^k_\alpha$ molecules, since its proliferation is supported by strain A and $[B10.A(3R) \times B10.A(4R)]F_1$ APC. Thus, clone 69 defines a fourth type of T_H cell clone involved in the response to Mb.

Analysis of the reactivity patterns of the above T_H cell clones suggests a framework for addressing a number of complex questions. All of the clones proliferate in response to Mb and a particular CNBr fragment. So far, we have isolated clones reactive with fragments 1-55 (clones 28.1 and 45.49) and 56-131 (clones 26.17 and 69). We have yet to isolate a clone which responds to a determinant on fragment 132-153, although reactivity to this fragment is present in a long-term T cell culture derived from whole lymph node cells. Perhaps this is because fragment 132-153 contains a "suppressor" epitope (12) which does not allow the isolation of the reactive clone by our methods.

Each of the clones isolated so far responds equally well to the appropriate CNBr fragment as it does to native Mb. As discussed above, this may be because T_H cells recognize "cross-reactive" determinants on both native and denatured forms of antigen. It is difficult to make absolute statements using T cell proliferation assays, however, since accessory cells (APC) are also necessary for antigen reactivity. In the case of clone 28.1, for instance, it is quite clear that the position of the dose response curve for Mb responsiveness is profoundly affected by the APC H-2 haplotype. What else can be learned about the mechanism of antigen presentation? We can conclude that effective antigen presentation does not require the entire Mb molecule, since CNBr fragments function equally well in stimulating appropriate clones. This suggests, for instance, that APC do not bind to residues 56-153 to present residues 1-55 to cloned T cells. This conclusion would seem to restrict the ways in which APC function in "determinant selection." In its most simple form, "determinant selection" suggests that a specific association occurs between antigen and APC Ia molecules so that a unique determinant is selected for T cell recognition (13). We should be able to rigorously test this theory. If clones 28.1 and 45.49, for instance, can be shown to recognize the same Mb determinant, we would conclude that two different Ia molecules function identically in antigen presentation of a single Mb determinant. We are currently attempting to define the precise determinants recognized by clones 28.1 and 45.49 using synthetically prepared peptide fragments of 1-55 and testing them for T-cell stimulation.

The quantitative response of clone 28.1 to both Mb and $I\text{-}A^b$ determinants also sheds light on APC function and its effects on T_H-cell immune responses. It suggests that T-cell dependent antigen responses are controlled by both qualitative ("all-or-none") and quantitative (graduated) mechanisms. Since the quantity of Ia molecules expressed on the surface of macrophages (which can function as APC) can be increased by treatment with lymphokines (14), this suggests that the expression of Ia might be a target *in vivo* for immune response control. Preliminary results suggest that gamma-interferon is important in regulating the level of Ia on APC (15). This may be a mechanism by which non-specific anti-inflammatory reactions can enhance specific immune responses.

Recently, we have also shown that Mb-reactive T_H cell lines function effectively to produce Mb-specific, anti-TNP antibody responses when added to TNP-KLH primed B cells and stimulated with TNP-Mb (Infante and Fathman, unpublished observations). An interesting finding of these studies is that anti-TNP antibody production induced by TNP-Mb is not necessarily associated with Mb-specific T-cell proliferation. Thus, antigen dosage seems important in providing distinct signals to the same T-cell population, i.e. high antigen dose signals T-cell proliferation and B cell help while the low doses provide B-cell help only. We have obtained evidence using KLH-reactive T-cell clones that these different responses involve two different B cell populations (Lyb 5^- and Lyb 5^+) (16). We are currently using cloned Mb and KLH reactive T cells as well as cloned APC (17) to determine how these two different antigen-induced signals are produced.

SUMMARY

In summary, we have sought to address a number of fundamental questions concerning the immune response to complex protein antigens by studying the responses of Mb-reactive T_H-cell clones. As discussed above, this approach can provide information relating to T-cell, APC and B-cell functions. We have taken a "reductionist" approach, in order to define minimal requirements for T-cell proliferation, Ir gene control, APC function and antibody production.

REFERENCES

1. Atassi, M.Z. Antigenic structure of myoglobin. The complete immunochemical anatomy of a protein and conclusions relating to antigenic structures of proteins. Immunochemistry 12:423, (1975).

2. Berzofsky, J.A., L.K. Richman and D.J. Killion. Distinct H-2 linked Ir genes control both antibody and T cell responses to different determinants on the same antigen, myoglobin. Proc. Natl. Acad. Sci. USA, 76:4046, (1979).

3. Okuda, K.P. Christadoss, S. Twining, M.Z. Atassi and C.S. David. Genetic control of immune response to lymphocyte proliferative response under H-2 linked Ir gene control. J. Immunol., 121: 866, (1978).

4. Kimoto, M. and C.G. Fathman. Antigen reactive T cell clones. I. Transcomplementing hybrid I-A-region gene products function effectively in antigen presentation. J. Exp. Med., 152:759, (1980).

5. Infante, A.J., M.Z. Atassi and C.G. Fathman. T cell clones reactive with sperm whale myoglobin. Isolation of clones with

specificity for individual determinants on myoglobin. J. Exp. Med., 154:1342, (1981).

6. Shigeta, M. and C.G. Fathman. I-region restrictions imposed upon the recognition of KLH by murine T cell clones. Immunogenetics, 14:415, (1981).

7. Atassi, M.Z. and B.J. Saplin. Immunochemistry of sperm whale myoglobin. I. The specific interaction of some tryptic peptides containing all the reactive regions of the antigen. Biochemistry, 1:688,(1968).

8. Endres, R.O. and H.M. Grey Antigen recognition by T cells. I. Suppressor T cells fail to recognize cross-reactivity between native and denatured ovalbumin. J. Immunol., 125:1515, (1980).

9. Infante, A.J. and C.G. Fathman. A T cell clone responds to quantitative differences in antigen and Ia on antigen presenting cells. (submitted for publication).

10. Berzofsky, J.A. Genetic control of the immune response to mammalian myoglobins in mice. I. More than one I-region gene in H-2 controls the antibody response. J. Immunol., 120:360, (1978).

11. Klein, J., A. Juretic, C.N. Baxevanis and Z.A. Nagy. The traditional and a new version of the mouse H-2 complex. Nature, 291:455, (1981).

12. Atassi, M.Z., S. Yokota, S.S. Twining, H. Lehmann and C.S. David. Genetic control of the immune response to myoglobin VI. Inter-site influences in T-lymphocyte proliferative response from analysis of cross-reactions of ten myoglobins in terms of substitutions in the antigenic sites and in environmental residues of the sites. Molecular Immunol. (in press).

13. Rosenthal, A.S., M.A. Barcinski, and J.T. Blake. Determinant selection is a macrophage dependent immune response gene function. Nature, 267:156, (1977).

14. Steeg, P.S., R.N. Moore, J.J. Oppenheim. Regulation of murine macrophage Ia antigen expression by products of activated spleen cells. J. Exp. Med., 152:1734, (1980).

15. McNicholas, J.M., D.B. Murphy, L.A. Matis, R.H. Schwartz, E.A. Lerner, C.A. Janeway, Jr., and P.P. Jones. Immune response gene function correlates with the expression of an Ia antigen. I. Preferential association of certain Ae and Eα chains results in a quantitative deficiency in expression of an Ae:Eα complex. J. Exp. Med. (in press).

16. Singer, A., Y. Asano, M. Shigeta, K.S. Hathcock, A. Ahmed, C.G. Fathman and R.J. Hodes. Distinct B cell subpopulations differ in their genetic requirements for activation by T helper cells. Immunol. Rev. (in press).

17. McKean, D.J., A.J. Infante, A. Nilson, M. Kimoto, C.G. Fathman, E. Walker and N. Warner. MHC restricted antigen presentation to antigen reactive T cells by B lymphocyte tumor lines. J. Exp. Med., 154:1419, (1981).

10

USE OF T CELL HYBRIDOMAS IN THE ANALYSIS OF HELPER T CELL FUNCTIONS IN B CELL RESPONSES

Neal Roehm, John Kappler and Philippa Marrack

Department of Medicine
National Jewish Hospital and Research Center
Denver, CO

INTRODUCTION

The fundamental question under investigation is to determine the role of helper T cells in the induction of B cell antibody synthesis. Based on our previous work we have proposed that helper T cells mediate two types of helper functions; the production of antigen non-specific factors i.e. lymphokines and antigen specific help[1,2].

One line of evidence supporting this model is a dichotomy in the helper activities required by B cells responding to red blood cell (RBC) associated antigens and soluble antigens such as keyhole limpet hemocyanin (KLH)[2]. Thus antibody responses against RBC associated antigens can be generated using either RBC primed T cells or supernatants of Con A stimulated spleen cells (Con A SN), a source of non-specific factors. In contrast the non-specific factors present in Con A SN are not sufficient for the B cell response to soluble antigens such as TNP-KLH.

A second line of evidence supporting the model is based on the use of antisera directed against I region antigens, prepared by immunizations with Con A blasts[2]. When carrier primed T cells with demonstrable helper function are depleted of Ia^+ cells by negative selection, the helper activity of these cells in the B cell response to soluble antigen is lost. However the addition of Con A SN, which is insufficient alone, to the carrier primed T cells depleted of Ia^+ cells, restores the antibody response. These results support the hypothesis that there are two functionally distinct helper T cells required for B cell responses to soluble antigen; an Ia^- T cell with antigen specific helper function and an Ia^+ T cell which produces non-specific factors, which are also found in Con A SN.

THE T CELL HYBRIDOMAS

One of the approaches we have pursued in analyzing T cell function is the formation of T cell hybridomas by the fusion of enriched antigen specific T cells to the T cell lymphoma BW5147[3]. The fusion products were screened for the ability to produce interleukin-2 (IL-2) in response to specific antigen and antigen presenting cells of the appropriate *H-2* haplotype. The production of IL-2 by the hybridomas was monitored by the ability of their culture supernatants to sustain the continued growth and viability of the IL-2 dependent T cell line HT-2 (kindly provided by J. Watson)[3].

The hybridomas relevant to this discussion are listed in Table 1, with their derivation and specificity. Each of the hybridomas produces IL-2 when stimulated with Con A or specific antigen and antigen presenting cells (normal irradiated spleen cells) of the appropriate *H-2* haplotype. Some related hybridomas with specificity for *I-A*^{k} plus chicken ovalbumin (OVA) also produce IL-2 when stimulated with *H-2*^{b} spleen cells in the absence of added antigen. This stimulatory activity maps to the *I-A* subregion.

It was originally thought that our success in preparing antigen specific hybridomas resulted from the use of the Con A inducible hybridoma FS6-14.13 and it's derivatives as fusion partners. However it has now become apparent that this is not the case and we have had equal success by directly fusing enriched antigen specific T cells to BW5147.

USE OF T CELL HYBRIDOMAS IN THE ANALYSIS OF T CELL HELPER FUNCTION

The assay system used in the experiments presented in this paper utilized modified Mishell-Dutton cultures[4,5]. The hapten-carrier conjugates used as antigen were trinitrophenol (TNP) coupled to various soluble protein carriers such as KLH, human gamma globulin (HGG) and OVA. The splenic B cell/macrophage pool was derived from mice primed one week earlier with TNP conjugated to lipopolysaccharide (LPS) and depleted of T cells by negative selection with monoclonal anti-Thy-1 and rabbit anti-mouse thymocyte serum. The helper T cells were derived from two sources; hybridoma cells irradiated with 4,000 R or carrier primed lymph node T cells prepared by passage over nylon fiber columns. Increasing numbers of these helper T cells were titrated into duplicate cultures and on day four of incubation the anti-TNP

Table I. Antigen Specific, *H-2*-restricted T Cell Hybridomas Used in These Studies

Hybridoma	Normal T Cell Parent: Strain	Normal T Cell Parent: Antigen Specificity	Fusion Partner	Hybridoma Produces IL-2 When stimulated with:
AODK 10.4	B10.D2 ($H\text{-}2^d$)	KLH	A040.10.AG1	Con A; KLH + $I\text{-}A^d$
AODH 3.4	DBA/2 ($H\text{-}2^d$)	HGG	A040.10.AG1	Con A; $H\text{-}2^b$; OVA + $I\text{-}A^k$
AODH 7.1	DBA/2 $H\text{-}2^d$)	HGG	A040.10.AG1	Con A; KLH + $I\text{-}E?^d$
BDK 11.1	$B6D2F_1$ ($H\text{-}2^{bxd}$)	KLH	BW5147[a]	Con A; KLH + $I\text{-}A^b$
A040.10.AG1	B10.A ($H\text{-}2^a$)	OVA	FS6-14.13.AG2	Con A; $H\text{-}2^b$; OVA + $I\text{-}A^k$
FS6-14.13	$B6D2F_1$ ($H\text{-}2^{bxd}$)	none	BW5147	Con A

[a]BW5147.G.1.4.0ua^r.1 kindly supplied by the Cell Distribution Center, Salk Institute.

direct plaque forming cell (PFC) response of the cultures is determined using the Jerne hemolytic plaque assay.

Because we were specifically interested in the antigen specific helper activity of the hybridomas, we tried to make sure that all other helper functions that could be required by B cells responding to TNP-proteins were saturating in our assays. This was done by adding excess Con A SN to the cultures as a source of non-specific factors. The Con A SN were prepared by stimulation of normal spleen cells with 4 µg/ml Con A for 24 hours and then harvesting the culture supernatants[6]. Alpha-methyl mannoside was added to the cultures at a concentration of 10 mg/ml to inhibit the activity of residual Con A. To monitor the effects of the non-specific factors present in Con A SN, control cultures received Con A media prepared in an identical manner to Con A SN except it was not incubated with spleen cells.

ANTIGEN SPECIFIC HELPER FUNCTION OF THE T CELL HYBRIDOMAS

The first experiment (Table 2) demonstrated the antigen specific helper function of two of the T cell hybridomas, AODK 10.4 specific for $\underline{H-2}^d$ plus KLH and AODH 7.1, specific for $\underline{H-2}^d$ plus HGG. When TNP-KLH was used as antigen both KLH primed lymph node T cells and the T cell hybridoma AODK 10.4 plus Con A SN allowed the generation of strong anti-TNP PFC responses. The PFC response generated with AODK 10.4 was totally dependent on the non-specific factors in Con A SN as responses were not generated using Con A media.

When the HGG specific hybridoma AODH 7.1 was assayed using TNP-KLH as antigen, significant anti-TNP PFC responses were not generated, however using TNP-HGG as antigen, significant antigen specific helper activity was apparent. Conversely, whereas AODK 10.4 was positive using TNP-KLH, antigen specific help was not observed using TNP-HGG as antigen with this hybridoma. In contrast to AODK 10.4, the anti-TNP PFC response generated with AODH 7.1 was not totally dependent on the non-specific factors present in Con A SN. Significant responses were generated using control Con A media. On this basis we would infer that AODH 7.1 produces non-specific factors besides IL-2 in addition to its antigen specific helper function and these substitute for Con A SN. Comparison of the PFC response generated with AODH 7.1 + Con A media with that generated using HGG primed T cells indicates AODH 7.1 had considerably greater helper activity than the carrier primed T cells on a per cell basis.

Table 2. Antigen Specific Helper Function of T Cell Hybridomas

Antigen: TNP-KLH			Antigen: TNP-HGG		
Helper T cell	x 10^{-5}	αTNP PFC per culture	Helper T cell	x 10^{-5}	αTNP PFC per culture
AODK 10.4	none	40	AODH 7.1	none	135
+	0.5	916	+	0.5	2,907
Con A SN	1.0	1,617	Con A SN	1.0	5,306
	2.0	2,780		2.0	6,720
AODK 10.4	none	40	AODH 7.1	none	25
+	0.5	40	+	0.5	1,450
Con A	1.0	30	Con A	1.0	2,730
media	2.0	60	media	2.0	3,160
AODH 7.1	none	135	AODK 10.4	none	70
+	0.5	45	+	0.5	55
Con A SN	1.0	135	Con A SN	1.0	40
	2.0	65		2.0	50
	4.0	110		4.0	90
KLH T cells	none	80	HGG T cells	none	90
	1.0	663		1.0	183
	2.0	1,067		2.0	443
	4.0	1,380		4.0	600
	8.0	2,687		8.0	1,237

Helper T cells in the form of carrier primed B10.D2 lymph node T cells or irradiated T cell hybridomas were added to duplicate Mishell-Dutton cultures of TNP-LPS primed B10.D2 B cells and macrophages. TNP-KLH or TNP-HGG were used as antigens. Con A SN, as a source of non-specific factors, or control Con A media were added to cultures with the T cell hybridomas. Anti-TNP PFC were measured on day 4 of culture using the Jerne hemolytic plaque assay.

H-2 RESTRICTION IN THE COLLABORATION OF T CELL HYBRIDOMAS AND B CELLS

The next issue we wished to investigate is the question of whether the delivery of the antigen specific helper signal to the B cell is H-2 restricted. The experiment again utilized a reciprocal design in which two KLH specific hybridomas, AODK 10.4 specific for KLH plus $H\text{-}2^d$ and BDK 11.1, which recognizes KLH plus $H\text{-}2^b$, were assayed for their ability to collaborate with $H\text{-}2^d$, B10.D2 B cells and $H\text{-}2^b$, B10 B cells (Table 3).

The premise of the experiment was that after irradiation with 4,000 R the ability of T depleted spleen cells to develop into PFC was blocked whereas their antigen presenting function remained intact. As demonstrated in Table 3, B10.D2 B cell/macrophages irradiated with 4,000 R were no longer able to develop into PFC, but functioned normally as antigen presenting cells in the production of IL-2 by AODK 10.4 upon stimulation with KLH (data not shown). Similarly, irradiated B10 B cell/macrophages were not able to develop into PFC, but functioned normally as antigen presenting cells in the production of IL-2 by BDK 11.1 stimulated with KLH (data not shown).

When AODK 10.4 ($H\text{-}2^d$ restricted) was added to an equal mixture of irradiated and non-irradiated $H\text{-}2^d$ B10.D2 B cell/macrophages good anti-TNP PFC responses were generated. In contrast, significant PFC responses were not generated using non-irradiated $H\text{-}2^b$ B10 B cells even though functional irradiated $H\text{-}2^d$ (B10.D2) antigen presenting cells were also added. The last group, irradiated B10 plus non-irradiated B10.D2 B cell/macrophages established that mixing these H-2 disparate populations did not inhibit the PFC response.

In the reciprocal component of the experiment the $H\text{-}2^b$ restricted hybridoma BDK 11.1 allowed the generation of an anti-TNP PFC response by non-irradiated $H\text{-}2^b$ B10 B cells, but not $H\text{-}2^d$ B10.D2 B cells, even in the presence of functional antigen presenting cells of the appropriate $H\text{-}2^b$ haplotype. The positive response generated using irradiated B10.D2 plus non-irradiated B10 B cell/macrophages again established this mixture of H-2 disparate cells did not inhibit the response.

We conclude on the basis of this and other experiments that the delivery of the antigen specific signal is restricted by H-2 antigens on at least the B cell.

REQUIREMENT FOR LINKED RECOGNITION

As described by Mitchison over a decade ago, antigen specific helper activity is typified by the requirement for linked recognition of the hapten-carrier conjugate[7]. Thus collaboration between hapten specific B cells and carrier primed T cells was not observed upon stimulation with free carrier and hapten conjugated to an unrelated carrier.

Table 3. H-2 Restriction in the Delivery of the Antigen Specific Helper Signal to the B Cell

AODK 10.4 x 10^{-5}	B Cell/Macrophage	Anti-TNP PFC	BDK 11.1 x 10^{-5}	B Cell/Macrophage	Anti-TNP PFC
none	irrad. B10.D2	0	none	irrad. B10	0
0.5		0	0.5		20
1.0		0	1.0		0
2.0		0	2.0		0
none	irrad. B10. D2	110	none	irrad. B10	220
0.5	+	490	0.5	+	890
1.0	B10.D2	1,590	1.0	B10	1,120
2.0		2,560	2.0		1,460
none	irrad. B10.D2	10	none	irrad. B10	30
0.5	+	90	0.5	+	0
1.0	B10	230	1.0	B10.D2	30
2.0		30	2.0		60
none	irrad. B10	20	none	irrad. B10.D2	110
0.5	+	680	0.5	+	820
1.0	B10.D2	1,380	1.0	B10	1,100
2.0		2,220	2.0		1,800

Irradiated KLH specific T cell hybridomas AODK 10.4 ($H-2^d$ restricted) or BDK 11.1 ($H-2^b$ restricted) were added to duplicate Mishell-Dutton cultures stimulated with TNP-KLH. Irradiated or non-irradiated splenic B cell/macrophage were taken from B10.D2 ($H-2^d$) or B10 ($H-2^b$) TNP-LPS primed mice.

Using an analogous approach we have tested the T cell hybridomas for the requirement for linked recognition in the delivery of antigen specific help. In the experiment presented in Table 4, AODH 7.1 (HGG specific) was assayed for its ability to allow the generation of a TNP specific PFC response following stimulation with TNP-HGG or free HGG plus TNP-KLH, the hapten conjugated to an unrelated carrier. The anti-TNP response generated using TNP-KLH and free HGG in the range of 2.0 to 500 µg/ml was less than 6% of the control response using TNP-HGG. The positive PFC response using KLH primed T cells and TNP-KLH as antigen established the immunogenicity of the TNP-KLH preparation.

Table 4. Requirement For Linked Recognition

Source T Cell Help	$\times 10^{-5}$	Anti-TNP PFC/culture µg/ml HGG none	2.0	10	100	500	Antigen
AODH 7.1	none	35					TNP-HGG
+	0.5	2,300					
Con A SN	1.0	2,990					
	2.0	4,080					
AODH 7.1	none	50	45	60	60	20	TNP-KLH
+	0.5	70	185	185	70	65	
Con A SN	1.0	20	100	225	235	25	
	2.0	20	60	220	230	35	
KLH T cells	none	5					TNP-KLH
	2.0	767					
	4.0	1,200					
	8.0	1,360					

B10.D2 KLH primed lymph node T cells or the HGG specific T cell hybridomas AODH 7.1 were added to duplicate cultures of TNP-LPS primed B10.D2 B cells. These cultures were stimulated with TNP-HGG or TNP-KLH +/- free HGG.

We have made similar observations using the KLH specific hybridoma AODK 10.4 and comparing the hapten specific PFC responses generated upon stimulation with TNP-KLH versus TNP-HGG and free KLH (data not shown). We would conclude that under conditions of limiting antigen, collaboration between hapten specific B cells and carrier specific T cell hybridomas requires linked recognition of the hapten-carrier conjugate.

CONDITIONS WHICH CIRCUMVENT THE REQUIREMENT FOR LINKED RECOGNITION

We have also tested the OVA specific T cell hybridoma AODH 3.4 for the requirement for linked recognition in its collaboration with TNP-LPS primed B cells of the $\underline{H\text{-}2}^a$ haplotype. In contrast to KLH and HGG, OVA can be added to Mishell-Dutton cultures at very high concentrations without inhibiting the generation of a PFC response.

In Table 5 we present the dose response profile of the anti-TNP PFC response generated using AODH 3.4 and varying concentrations of TNP-OVA or OVA alone. As demonstrated in this experiment optimal anti-TNP PFC responses using TNP-OVA occur at low antigen concentrations in the range of 1-10 µg/ml. Using OVA alone in the absence of TNP, equivalent hapten specific PFC responses can also be generated at very high OVA concentrations on order of 1-5 mg/ml.

While this experiment appears to contradict our previous conclusion regarding requirements for linked recognition, there is a crucial difference. At high concentrations HGG and KLH inhibit PFC responses whereas OVA does not. Previous studies from this laboratory[8] suggest that B cells can take up exogeneous antigen non-specifically, "process" it, and present it to helper T cells in association with their surface Ia. More recent experiments from our[9] and other laboratories have confirmed this idea using B cell tumor lines. Presumably very high antigen concentrations are needed to drive this process in B cells.

It is our hypothesis that the presence of a TNP specific immunoglobulin receptor allows the B cells to bind TNP-carrier efficiently, while in the absence of a specific antigen receptor this uptake is inefficient and high concentrations of antigen are required. We would further hypothesize that recognition of OVA + $\underline{I\text{-}A}^k$ on the B cell surface by AODH 3.4 caused direct activation of the hapten primed B cells without the apparent involvement of the B cells immunoglobulin receptor. On the basis of this idea we wondered if direct recognition of the B cells $\underline{I\text{-}A}$ antigen could lead to B cell activation.

As mentioned above, AODH 3.4 recognizes $\underline{I\text{-}A}^k$ plus OVA or $\underline{I\text{-}A}^b$ alone. We have therefore tested AODH 3.4 for its ability to collaborate with TNP-LPS primed B cells of various $\underline{H\text{-}2}$ haplotypes including $\underline{H\text{-}2}^b$ in the generation of an anti-TNP PFC response in the apparent absence of antigen. As shown in Table 6 the generation of significant hapten specific PFC responses only occurred using B

Table 5. Absence of Linked Recognition with High Antigen Concentration

T Cell Hybridoma	x 10^{-5}	Antigen	αTNP-PFC per culture
AODH 3.4 + Con A SN	none	none	90
	0.5		90
	1.0		50
	2.0		130
	none	0.1 μg/ml TNP-OVA	45
	0.5		140
	1.0		320
	2.0		910
	none	1.0 μg/ml TNP-OVA	5
	0.5		325
	1.0		710
	2.0		1,190
	none	10 μg/ml TNP-OVA	5
	0.5		195
	1.0		545
	2.0		1,080
	none	10 μg/ml OVA	20
	0.5		25
	1.0		60
	2.0		180
	none	100 μg/ml OVA	75
	0.5		85
	1.0		120
	2.0		230
	none	1.0 mg/ml OVA	40
	0.5		145
	1.0		280
	2.0		700
	none	5 mg/ml OVA	35
	0.5		560
	1.0		980
	2.0		1,210

cells of the $\underline{H\text{-}2}^b$ haplotype and not with B cells of the $\underline{H\text{-}2}^d$, $\underline{H\text{-}2}^f$ or $\underline{H\text{-}2}^q$ haplotypes.

As a control demonstrating the ability of the B cells to generate anti-TNP PFC responses, duplicate cultures of the B cells were stimulated with TNP-OVA and OVA primed T cells syngeneic to the B cells. In every case satisfactory responses were obtained (Table 6).

Table 6. Hybridoma-B Cell Collaboration Without Antigen

No antigen				Antigen: TNP-OVA		
B Cell Source		$\times 10^{-5}$	αTNP PFC per culture	B Cell Source	$\times 10^{-5}$	αTNP PFC per culture
B10	AODH 4.3	none	10	B10	none	25
($\underline{H\text{-}2}^b$)	+	0.5	640	OVA	2.0	150
	Con A SN	1.0	800	T cells	4.0	280
		2.0	1,010		8.0	320
B10.D2	AODH 3.4	none	5	B10.D2	none	10
($\underline{H\text{-}2}^d$)	+	0.5	10	OVA	2.0	165
	Con A SN	1.0	0	T cells	4.0	305
		2.0	20		8.0	430
B10.M	AODH 3.4	none	20	B10.M	none	15
($\underline{H\text{-}2}^f$)	+	0.5	0	OVA	2.0	170
	Con A SN	1.0	50	T cells	4.0	345
		2.0	15		8.0	510
B10.Q	AODH 3.4	none	25	B10.Q	none	10
($\underline{H\text{-}2}^q$)						
	+	0.5	10	OVA	2.0	130
	Con A SN	1.0	20	T cells	4.0	275
		2.0	10		8.0	405

AODH 3.4 was added to TNP-LPS primed B cells from B10 ($\underline{H\text{-}2}^b$), B10.D2 ($\underline{H\text{-}2}^d$), B10.M ($\underline{H\text{-}2}^f$) or B10.Q ($\underline{H\text{-}2}^q$) mice without the addition of antigen. As a control these same B cell populations were also tested with syngeneic OVA primed lymph node T cells and TNP-OVA as antigen.

Further experiments established that AODH 3.4 only stimulated TNP-primed $\underline{H\text{-}2}^b$ B cells. $\underline{H\text{-}2}^b$ B cells from normal mice or mice primed with LPS alone were not activated in this system. We would infer that stimulation of B cells by T cell recognition of surface I-A only by passed antigen-specific signals to the B cell if the latter cell were already "activated" in some fashion, in this case by prior antigen stimulation. The other possibility, that small amounts of TNP-LPS were carried over into our cultures from the primed animal, also still stands and must be tested.

In summary the T cell hybridoma AODH 3.4 demonstrated antigen specific helper function using $H\text{-}2^a$ B cells stimulated in culture with low doses of TNP-OVA or high doses of OVA alone. AODH 3.4 also allowed the generation of a hapten specific PFC response by TNP-LPS primed $H\text{-}2^b$ B cells in the absence of overtly added antigen.

FURTHER SPECULATIONS OF THE NATURE OF ANTIGEN SPECIFIC HELPER T CELLS

At this point in our studies we have screened six randomly chosen independent T cell hybridomas for antigen specific helper function. These hybridomas were originally selected solely on the basis of their ability to produce IL-2 in an antigen specific, H-2 restricted manner. Surprisingly each of these six hybridomas demonstrated antigen specific helper function in an assay system which was deliberately rigged to be independent of the production of non-specific factors by the addition of Con A SN. We are very much intrigued by the correlation between the ability of the hybridomas to recognize antigen in association with I region products and their ability to mediate antigen specific help, H-2 restricted. Based on this correlation we are inclined to speculate that for antigen specific helper activity T cells must simply have a receptor for antigen plus I, and that no other functions are required, provided sufficient nonspecific helper factors (Con A SN) are present.

SUMMARY

Using the T cell hybridomas we have presented data demonstrating the following points:

1. T cell hybridomas selected by their ability to produce IL-2 in an antigen specific, H-2 restricted manner also have carrier specific helper function in the B cell response to hapten, independent of their ability to make IL-2.

2. Delivery of the antigen specific helper signal is restricted at least by H-2 antigens on the B cell.

3. Under conditions of limiting antigen the T cell hybridoma's helper activity requires linked recognition.

4. Activation of hapten primed B cells to antibody synthesis can occur via an apparently immunoglobulin independent mechanism;

   a) Using very high concentrations of an antigen which is recognized by the T cell hybridoma.

   b) When the T cell hybridoma directly recognizes the B cell I-A antigen.

ACKNOWLEDGMENTS

The authors would like to thank Jim Liebson, Janice White and Ella Kushnir for their excellent technical assistance and Edna Squillante for the preparation of this manuscript. This work was supported by grants AI-17134 from the U.S. Public Health Service, grant IM-49 and FRA-218 from the American Cancer Society, Inc.

REFERENCES

1. J. E. Swierkosz, P. Marrack and J. W. Kappler, Functional analysis of T cells expressing I-A antigens. I. Demonstration of helper T cell heterogeneity, J. Exp. Med. 150:1293 (1979)

2. D. M. Keller, J. E. Swierkosz, P. Marrack and J. W. Kappler, Two types of functionally distinct, syngergizing helper T cells, J. Immunol. 124:1350 (1980)

3. J. W. Kappler, B. Skidmore, J. White and P. Marrack, Antigen-inducible, H-2-restricted, Interleukin-2-producing T cell hybridomas, J. Exp. Med. 153:1198 (1981)

4. P.C. Marrack and J. W. Kappler, The role of H-2 linked genes in helper T-cell function. III. Expression of immune response genes for trinitrophenyl conjugates oif Poly-L(Tyr,Glu)-Poly-D-L-Ala--Poly-L-Lys in B cells and macrophages, J. Exp. Med. 147:1596 (1978)

5. R. I. Mishell and R. W. Dutton, Immunization of dissociated spleen cell cultures from normal mice, J. Exp. Med. 126:423 (1967)

6. L. Harwell, J. W. Kappler and P. Marrack, Antigen-specific and nonspecific mediators of T cell/B cell cooperation. III. Characterization of the nonspecific mediator(s) from different sources. J. Immunol. 116:1379 (1976)

7. N. A. Mitchison, The carrier effect in the secondary response to hapten-protein conjugates. II. Cellular cooperation, Eur. J. Immunol. 1:18 (1971)

8. P. Marrack and J. W. Kappler, The role of H-2-linked genes in helper T cell function. VII. Expression of I region and immune response genes by B cells in bystander helper assays, J. Exp. Med. 152:1274 (1980)

9. E. Walker, N. Warner, R. Chesnut, H. M. Grey, P. Marrack and J. W. Kappler, manuscript in preparation.

10. D. McKean, A. J. Infante, A. Nilson, M. Kimoto, C. G. Fathman, E. Walker and N. Warner, MHC restricted antigen presentation to antigen reactive T cells by B lymphocyte tumor cells, J. Exp. Med. in press (1981)

11

REGULATION OF GROWTH AND DIFFERENTIATION IN B CELL CLONES AND HYBRIDS

David W. Scott, P. S. Pillai, Scott J. Anderson
and Ronald B. Corley

Division of Immunology, Duke University Medical Center,
Durham, N. C. 27710

Analysis of the molecular events which occur during specific B cell activation or inactivation has been hampered by the rare frequency of antigen-specific B cells, as well as by the heterogeneity of B cell subpopulations and stages of differentiation. An initial approach to the solution of this dilemma is to enrich for antigen-specific lymphocytes. The purification of B cells of a single hapten-specificity has, indeed, been possible for a number of years (1). However, such enrichment procedures do not overcome the heterogeneity which exists in B cells of a single specificity. The ability to clone hapten-specific B cells and/or hybridize them to an appropriate tumor cell would resolve many of these difficulties, assuming that the "normal" B cell phenotype can be retained. In analogy to the technology of T cell cloning, which is abundantly apparent at the present meeting, clones of hapten specific B cells are now being prepared. In contrast to T cell cloning, however, it is not yet possible to repeatedly stimulate B cells with specific antigen in the absence of terminal differentiation to antibody secreting cells. The present report describes the current state of the art with regard to B cell cloning technology to yield functional short-term hapten-specific clones as well as long-term hapten-specific inducible hybrids. We will focus on the kinds of B cell clones which we are able to grow and detect in terms of differential function and will begin an analysis of the regulation of the growth and/or differentiation of such clones by antigen, mitogenic growth potentiators, and cell-derived factors. Finally, the use of cell hybridization technology will be described. This yielded several B cell tumor hybrids, one of which has been studied in detail. The regulation of the growth of such a hybrid by T cell products will be introduced.

MATERIALS AND METHODS

Animals

$B6D2F_1$ (C57BL/6J x DBA/2J), CAF_1 (BALB/cJ x A/J) and BALB/cByJ mice from Jackson Laboratories were used at 6-12 weeks of age in all experiments. Rat thymus filler cells, from weanling female Lewis rat donors, were used in some experiments as indicated below.

Antigen

Fluorescein (FL) was conjugated to a number of thymus independent or thymus-dependent antigens as described earlier (2). The following carriers were employed: polymerized flagellin (POL), *E. coli* lipopolysaccharide (LPS), purified protein derivative (PPD), killed *Brucella abortus* organisms (BrA), and keyhole limpet hemocyanin (KLH).

Helper T Cell Populations

Highly enriched KLH-specific helper T cells were prepared by modifications of a previously described method (3,4). Briefly, draining lymph node cells from mice immunized with KLH according to the method of Corradin et al. (5) were enriched in T cells by passage over nylon wool columns and cultured with irradiated syngeneic spleen cells and KLH (3,4). Activated T cell blasts were selectively recovered at weekly intervals and recultured with KLH and irradiated syngeneic spleen cells (Corley, R. B., in preparation). Helper T cells were used between 2 and 5 weeks of *in vitro* culture. Independent experiments have established that such populations have excellent helper activity for both primed (3) and unprimed (4) B cells *in vitro* (Pillai, P. S., D. W. Scott, and R. B. Corley, in preparation).

B Lymphocyte Cloning in Soft Agar

The methods described in our earlier publications (6,7), which are modifications of the original procedures described by Metcalf et al. (8) and Kincade et al. (9), were employed routinely (Fig.1). In all cases, the B cells were first purified via the FL-gelatin procedure described by us (10) as modified from the orginal method of Haas and Layton (1). Hapten-enriched B cells were able to grow in this soft agar system with a linear cell dose response curve in response to mitogens or other potentiators; the cloning efficiency was usually 1 to 2% of the plated B cells (6). These data are consistent with those described for unpurified B cells by Kincade (11).

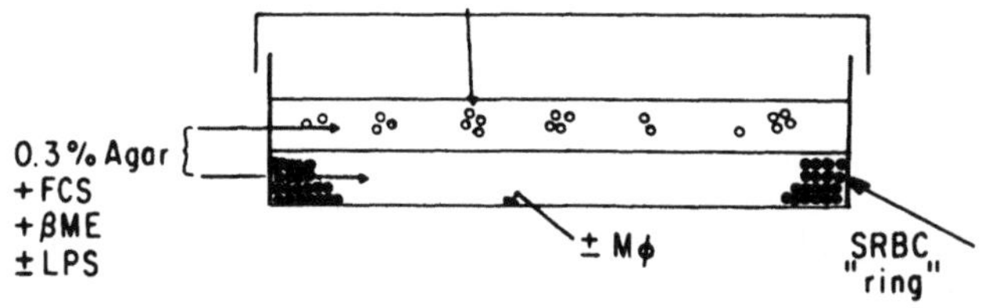

Figure 1. Schematic view of the method for cloning hapten-specific B cells in agar. The SRBC "ring" is initially placed at the periphery of the culture well as described earlier (6).

Microculture of B Cell Clones

Progeny cell colonies derived from a single hapten-purified B cell were manually picked from the agar after 3 to 7 days. Each colony was cultured in a single microtiter well in a volume of 200 microliters containing 1 of several FL-antigens plus either rat thymus filler cells, irradiated syngeneic spleen cells, or irradiated syngeneic spleen cells and irradiated KLH-primed helper T cell populations. A minimum of 32 to 48 discrete clones were cultured individually for each antigen in any given experiment.

Somatic Cell Hybridization

Two approaches using somatic cell hybridization were employed in order to create "immortalized" hapten-specific B cell lines. The first method involved the fusion of freshly prepared FL-gelatin enriched B cells with a variety of B lymphoma lines carrying a HAT-sensitivity marker (12, Anderson et al., in preparation). Preliminary experiments established the feasibility of this approach: that is, hapten-specific B cells can be fused with the BCL_1 lymphoma tumor line to yield heterokaryons which can be stimulated by hapten on LPS (12). Since the BCL_1 parent is responsive to LPS, the result of the stimulation of heterokaryons is the differentiation into anti-FL secreting plaque forming cells. Thus, the fluorescein specificity can be introduced into a B cell tumor line of unknown specificity (12). The procedure of Gefter et al. (13), who utilized polyethylene glycol (PEG) cell fusion in suspension, was used in all cases.

The second approach to produce hapten-specific immortalized lines involved the fusion of B lymphoma lines with an antibody-secreting hybridoma specific for fluorescein (the product of the fusion of MPC 11 myeloma cells with an FL-primed spleen population). In this procedure, the hybridoma secreting anti-fluorescein is substituted for normal FL-specific B cells. Such a fusion resulted in a number of hybrids (that is, double hybridomas) which were antibody-secreting as well as hybrids which were apparently non-secreting, but were still hapten-specific (by rosetting). Wells which yielded clones phenotypically falling into the latter category were expanded and re-cloned for further study. In the experiments described in this report, the following tumors were employed: 70Z/3, a pre-B cell line from B6D2 mice described by Paige et al. (14), BCL_1, an LPS-responsive lymphoma of BALB/c mice described by Slavin and Strober (15), and WEHI 231, also an LPS responsive B cell line (16), originally isolated and provided by Dr. Noel Warner.

RESULTS

Growth of Hapten-Specific B Cell Clones in Soft Agar

When FL-enriched B cells are plated in soft agar and grown for 6-7 days, typical B cell colonies (CFU-B) are observed (Fig. 2A-D). As noted earlier, the growth of these cells is mitogen dependent (6, 9, 11). Moreover, morphological differences between colonies grown under different conditions can often be discerned. LPS-stimulated colonies are often dispersed, whereas SRBC-potentiated CFU-B are more compact; each of these kinds of colonies are seen with both additives. The possibility that these morphologically different colonies may reflect functionally heterogeneous B cell subsets with distinct activation requirements will be discussed below.

It should be reiterated that these colonies are the progeny of a single B cell and that we know the specificity of the B cell. The hapten-specificity of these populations has been verified repeatedly (6), and the surface isotype demonstrated to be representative of normal B cells taken from animals of similar age (10). Moreover, upon growth under agar cloning conditions <u>in vitro</u>, it can be shown that approximately 85% of the B cells specifically bind fluorescein, and that this binding is inhibitable by anti-μ (M. McGrath, personal communication). Since each colony represents the progeny of a single B cell, it is possible to investigate the functional potential of such cells upon isolation from the agar matrix. We have previously reported that such hapten-specific CFU-B can be stimulated by several thymus-independent FL-conjugated antigens at high frequency (1/3 to 1/10), although the PFC per positive clone may not be large (6). A summary of those results is presented in Table 1A. More recently, we have established that these B cell colonies can also be triggered by the thymus-dependent antigen FL-KLH in the presence of carrier-specific T helper cell populations plus syngeneic antigen-presenting cells (Table 1B).

During the course of these studies, it was important to establish the parameters of growth of these B cell colonies and in so doing we observed that discrete colonies are clearly detectable as early as day 3 after plating. Some of these colonies can be followed on gridded plates and observed to be expanded on day 6-7. On the other hand, a fraction of these colonies (25-30%) disappear by day 7. A final population of CFU-B are those which are not present on day 3 but clearly detectable on day 6-7. Interestingly, when the early colonies are picked and stimulated with either TI or TD antigens, only the TD antigen will trigger differentiation to PFC. In one experiment, the majority of the 3 day colonies were triggered by TD antigen and none by FL-POL (Pillai, Scott and Corley, in preparation). The late colonies, as noted above, can be triggered by both TD and TI antigens. We are presently trying to formally establish whether the colonies noted on gridded plates on day 3 and still present on day 7 are restricted to differentiate only with TD antigenic stimulus. At the present time, there is only the functional correlation.

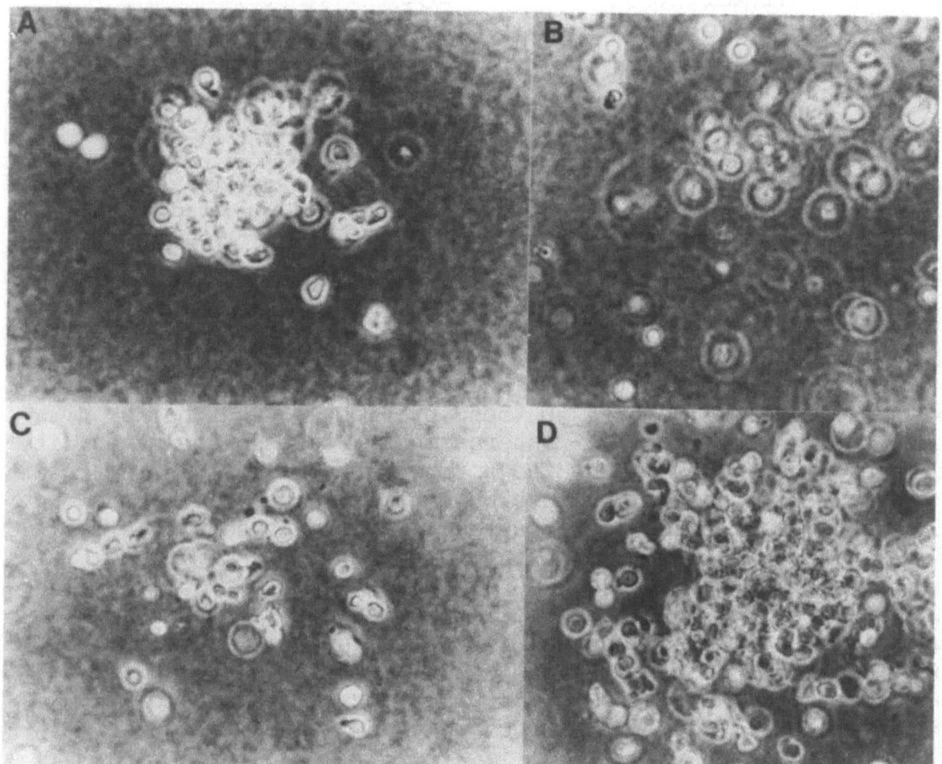

Figure 2. Hapten-specific B cell clones. CFU-B grown in the presence of SRBC plus LPS for 6-7 days show variations in clonal morphology. A) Compact cluster typical of clones also seen with SRBC alone. B and C) Dispersed colonies more commonly observed with CFU-B grown with LPS. D) Large colony composed of greater than 100 hapten-specific B cells.

Table 1. Restimulation of FL-Specific CFU-B by TI and TD Antigen*

Antigen	% Positive Clones†
Exp. A	
FL-POL	14.7
FL-LPS	20.1
FL-BrA	33.9
FL-PPD	8.3
Exp. B	
FL-POL	18.6
FL-KLH	29.6

*FL-specific B cells were cloned in soft agar in the presence of SRBC (Exp. A) or SRBC + LPS (Exp. B) and CFU-B progeny picked on d7 and restimulated with either FL-POL or FL-KLH, the latter in the presence of 10^5 irradiated KLH-specific helper cells and 10^5 irradiated syngeneic filler spleen cells.

†7-25 mean PFC per positive clone.

Earlier we reported that mitogen-potentiated hapten-specific CFU-B are similar in numbers to those previously described by Kincade and co-workers (11). Interestingly, the subsequent ability to trigger these progeny CFU-B to antibody secretion by FL-antigens is related to the initial culture conditions. Thus, cells grown in the presence of sheep erythrocytes were stimulatable by all TI antigens we have employed, albeit to varying degrees, whereas those cells grown with LPS but no sheep erythrocytes could not be triggered by any TI antigens. This unresponsiveness to antigenic triggering was not due solely to terminal differentiation induced by LPS since cultures grown in the presence of both SRBC and LPS were still triggerable (6). Further experiments are necessary to determine if LPS grown colonies can be triggered by TD antigens.

Effect of Tolerogen on CFU-B Growth and Differentiation

The data presented thus far demonstrate that hapten-specific B cells can be purified, cloned in soft agar, and subsequently stimulated by a variety of thymus-independent and thymus-dependent

Table 2. Effect of Tolerogen on Growth and Antigen-Induced Differentiation of FL-Specific CFU-B*

Tolerogen	CFU-B/10^4	% Positive Clones
+	206	7.8
-	373	15.8

*FL-specific B cells were grown with SRBC + LPS in the presence (+) or absence (-) of 25μg FL-SGG, picked and challenged with FL-POL in microculture as described earlier.

antigens (the latter in the presence of T cell help). Our data also suggested that the presence of different potentiators (mitogens) alone or in combination may stimulate the growth of functionally different B cell subsets. Because these B cell clones retain immunologic competence and their specificity is known, it was possible to look at the effect of tolerogen on the growth and subsequent stimulation of the progeny of a single B cell. Two approaches have been employed, both with similar results. FL-specific B cells are cloned in soft agar in the presence of FL-SGG or are pre-incubated with FL-SGG and then washed before agar cloning. Since greater than 75% of the initial population can be shown to directly bind antigen, an effect of tolerogen interacting via specific receptors could be expected. Indeed, we found that a portion of the CFU-B are inhibited in their growth in the presence of tolerogen (Table 2). This inhibition varied from 25-50% and was maximal at 25μg FL_6SGG per dish (7). Thus the clonal expansion of specific B cells could be inhibited by the presence of tolerogen. This effect was specific and due to interaction of the fluorescein moiety because TNP-SGG or unhaptenated SGG did not inhibit B cell growth (7). Fluorescein-ficoll and monovalent fluorescein were ineffective, a result which suggests that the gamma globulin carrier is "tolerogenic"in this system,perhaps by interacting via an F_c receptor on B cells.

The remaining colonies,which grew in the presence of tolerogen, were tested for their ability to be triggered by FL-POL into plaque forming cells. The data presented in Table 2 suggest that the surviving CFU-B progeny, while still able to divide in the presence of tolerogen, could not be triggered to differentiate into PFC by our challenge antigen, FL-POL. Thus, two effects of tolerogen have been established: first, a portion of CFU-B fail to grow in the presence of tolerogen, indicating that this antigenic interaction affects a subpopulation in its clonal expansion. Secondly, the majority of the remaining CFU-B which do grow from tolerogen no longer can be triggered to form PFC in the presence of antigen. This is a more stringent test for unresponsiveness.

This sensitivity to tolerogen by adult B cells is interesting since Nossal et al. recently reported that CFU-B grown from hapten-specific B cells obtained from adult spleens were insensitive to tolerogenic signals, whereas purified cells from neonatal sources could be tolerized by brief exposure to tolerogen in vitro (17). Since these workers grew their B cells in the presence of sheep red cells, LPS, and macrophages to optimize growth, we sought to establish whether CFU-grown under different conditions showed a differential susceptibility to tolerogen. Indeed, our initial studies established that CFU-B grown with SRBC or LPS alone could not be inhibited in their clonal growth by the presence of tolerogen; in the same experiment, a 30-40% inhibition of SRBC plus LPS grown colonies was observed (7) (Fig. 3). Thus, only those colonies grown with both additives contain a subset which is sensitive to tolerogen. Since the presence of macrophages will not lead to an

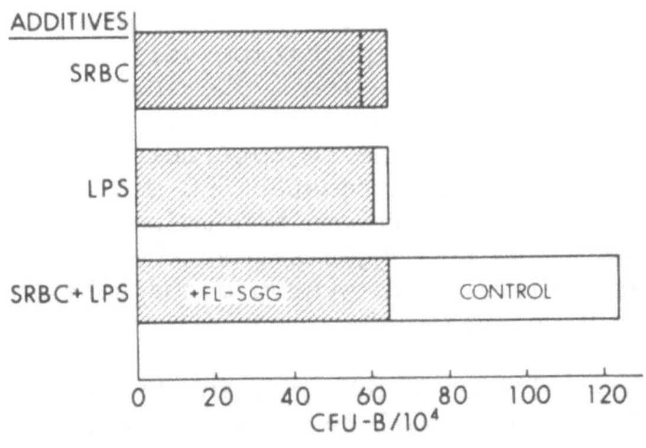

Figure 3. Effect of tolerogen on hapten specific CFU-B growth. FL-SGG at 25 μg per dish (stripped bars) or no antigen (control, open bars) was added to the agar cloning matrix in the presence of SRBC, LPS, or both additives. Only CFU-B grown in the presence of both additives were inhibited in their clonal expansion by tolerogen (bottom bar).

increased number of B cell colonies, but merely modulate the size of clones (6,11), it seemed possible that their presence during the agar cloning phase might reverse the effect on proliferation induced by tolerogen. The data presented in Table 3 support this notion and also help explain the discrepancy between Nossal's (17) results and ours. For example, B cells grown in the presence of SRBC and LPS are inhibited approximately 40% in CFU-B numbers, whereas the same FL-hapten specific B cells grown with SRBC, LPS and macrophages are insignificantly affected. Interestingly, however, a negative "signal" has still been received by the majority of the B cells since the subsequent triggering with FL-POL is reduced from one out of 5 to one out of 30 clones. Thus, macrophages have apparently reversed a negative signal with regard to B cell growth, but were unable to modulate the unresponsiveness with respect to differentiation induced by antigen.

The ability to follow the clonal expansion of the progeny of single B cells, and to define B cells functionally with respect to antigenic stimulation, regulation by tolerogen, and modulation of growth by accessory cells, will hopefully allow for a more detailed analysis of the mechanism of B cell triggering and tolerance now analyzed at the clonal level.

Table 3. Role of Macrophages in the Modulation of CFU-B Growth and Differentiation

Potentiators*	Tolerogen	CFU-B/10^4	% Positive Clones
SRBC + LPS	+	161	3.5
	-	286	17.8
SRBC + LPS + MΦ	+	218	3.5
	-	220	21.6

*FL-specific B cells were grown with SRBC + LPS ± MΦ in the presence or absence of FL-SGG, picked and challenged as described.

Functional Antigen-Specific B Cell Hybrids

An obvious goal of our current work has been to establish functional long term antigen-specific B cell lines. The data presented in previous sections demonstrate that antigen-specific and functional B cells can be cloned and their progeny studied. However, such cells could not be maintained beyond several weeks *in vitro* and we have not been able to further expand such populations in the presence of lymphokine containing supernatants. However, some success in the cloning of hapten-purified B cells using IL-1 and/or IL-2 has been achieved in collaboration with Maureen Howard and Bill Paul at the NIH (18). These will not be detailed here. Instead, an alternate approach which we have used is to hybridize B lymphoma lines with hapten-specific B cells with the aim of providing a known hapten specificity to a B cell tumor line with some functional capabilities. The feasibility of this approach with BCL_1 has already been reported (12). However, as noted in Table 4, no functional long-term antigen-reactive B hybridoma lines have been obtained. So far, over 30 cloned hybrids obtained from fusions by 70Z/3, WEHI-231, and BCL_1 have been studied: these appear to bind antigen but have not been triggerable with a variety of antigens *in vitro* or *in vivo*. Limited mitogen responsiveness of the

Table 4. B Cell Tumor x FL-SC Hybrids

Tumor Parent*	Number of Hybrid Clones	RFC	PFC†
WEHI 231B	10	+	-
70Z/3B	21	+	-
BCL_1**	1	-	-

*Several fusions with each tumor and FL-SC were performed, using 30-40% PEG, resulting in the indicated number of hybrid clones.

**After each BCL_1 x FL-SC fusion, the fused cells were grown from 1-7 days, then repassed over FL-gelatin. Adhering cells were injected i.v. into sublethally irradiated mice.

†There was very little variation among the hybrids derived from a single tumor parent; antigens used for triggering included FL-POL, FL-LPS, FL-BrA or FL-RBC. No significant change in ^{3}HdT incorporation was induced by antigen in a hapten-specific manner.

parent lymphoma cell is often retained but no hapten-specific effects have been noted. Moreover, the capacity to rosette with antigen-coated red cells has often been lost in these cells despite the use of cloned populations. Perhaps this reflects the instability of such hybrids.

Fusion of a FL-Specific Hybridoma with B Lymphoma Lines

An alternate approach which has proved useful is the fusion of a hybridoma secreting anti-FL antibody of the IgM class with B lymphoma lines. This fusion, especially with the 70Z/3 pre-B cell parent, has yielded a number of lines and has been studied extensively. One such line, termed F9xZ-3, has been cloned 4 times and shows a very high percentage of rosette forming cells, with a small minority (less than 1%) of antibody secreting cells. In the presence of haptenated-LPS, or LPS alone,no increase in proliferation of these hybrids has been observed. However, some of these cells can be triggered to secrete anti-fluorescein antibody (Fig.4); as much as a ten fold increase in the number of antibody secreting cells can be induced within 2-3 days. While the absolute numbers of PFC induced by LPS or FL-LPS do not differ, the PFC detected when FL-LPS is used appear larger (Fig. 5). It is noteworthy that the best stimulation is obtained at 100 micrograms per ml FL-LPS, an unusually high dose, although a significant increase in antibody secretion can be induced with nanogram amounts of FL-LPS. Activation occurs in the absence of added accessory cells, and in a variety of culture media (Anderson and Scott, in preparation). Cell sorter analysis has revealed that the most readily activated cells within this cloned population are large, high density μ-bearing cells. Efforts to synchronize this population to determine the stage of cell cycle, if any, during which activation to antibody secretion is most effective, have not been successful as yet. Cytofluorograph analysis also shows that the population binding FL-LPS is inhibited by an anti-μ reagent. (Fig. 6). However, whether an interaction of the IgM receptor on these clones with hapten is necessary for activation by FL-LPS can not yet be formally determined.

Response to T Helper Cells

During the course of this study we investigated whether such a hybrid line could be triggered by T-dependent antigens in the presence of appropriate helper cells. The same KLH-specific helper cells employed earlier were utilized in this study. Surprisingly, no significant induction of anti-FL secretion was induced by FL-KLH plus KLH primed helper cells and accessory cells, even though such populations of helper cells are highly efficient in inducing normal B cells or hapten-specific CFU-B colonies to antibody secretion. However, we observed that our hybrid line was not insensitive to signals from helper T cells. That is, a

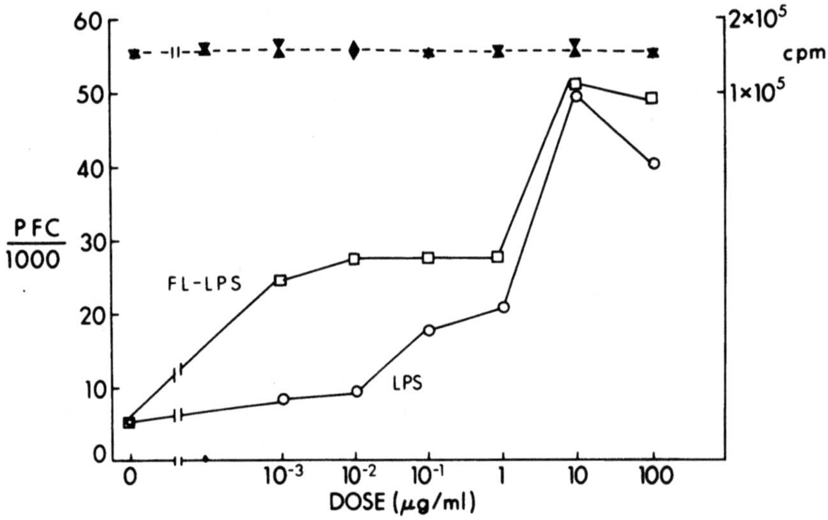

Figure 4. Induction of anti-FL secretion in the F9xZ (clone 3.11) hybridoma by FL-LPS (squares) or unhaptenated LPS (circles). Cells were cultured at 10^5 per 1 ml DMEM-FCS and samples examined on day 3 for the number of PFC/1000 cells. Aliquots were also labeled with ^3H thymidine and assayed for incorporation 18 hours later (inverted triangles, LPS; triangles, FL-LPS).

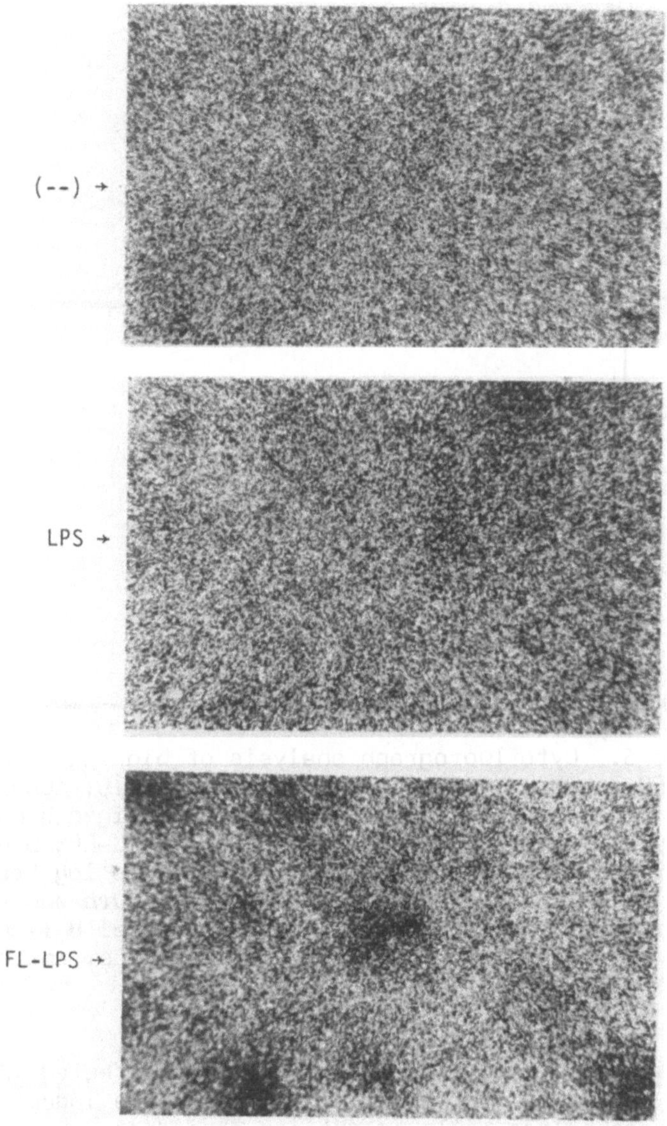

Figure 5. PFC morphology of F9xZ cells cultured alone with LPS or with FL-LPS, as in the previous figure. Note the larger PFC induced by FL-LPS.

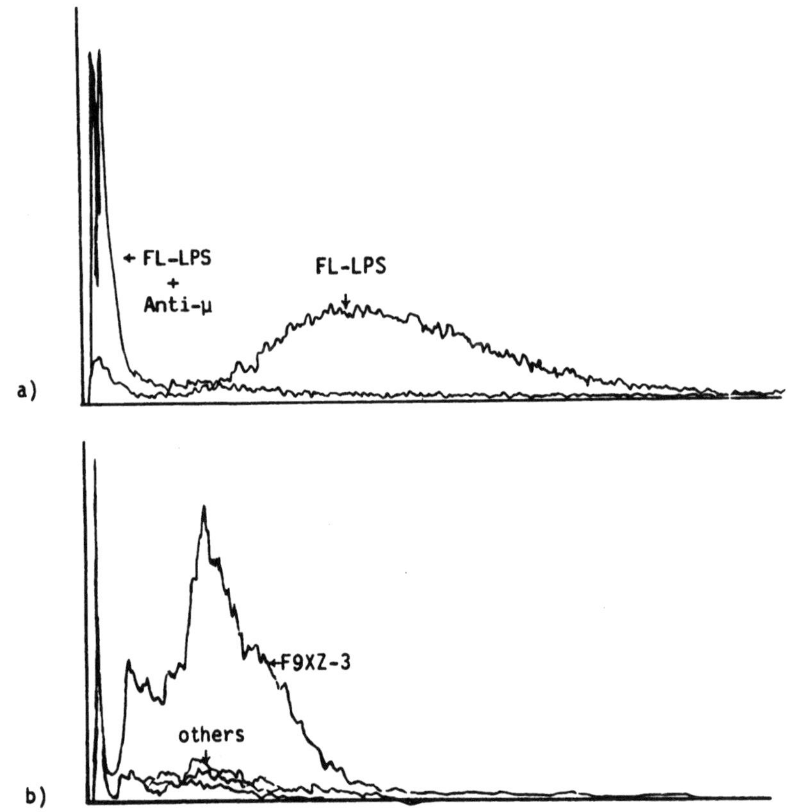

Figure 6. Cytofluorograph analysis of binding of FL-LPS by the F9xZ hybrid. Ordinate, number of cells; abscissa, fluorescence intensity. A) Inhibition of antigen binding by anti-μ. B) Independent confirmation of FL-LPS binding by the F9xZ hybrid. Fluorescence intensity is log transformed. Other cell lines were examined in this figure and found negative for binding. These included several B lymphoma lines of unknown specificity, a myeloma line, and the 70Z/3 parental pre-B cell.

dramatic increase in proliferation could be effected in the F9xZ line by T helper cells (Table 5). Although the induction of antibody by these helper cells is antigen-dependent, no apparent dependence on antigen was observed during these experiments, possibly because the F9xZ line is highly sensitive to growth-promoting factors released by activated helper cells. Present studies are directed toward an analysis of hybrids of these T helper cells and their products, which may be able to modulate the growth of cloned antigen-specific hybridoma lines. It should be noted that the effect on hybridoma growth by T helper cells was not restricted

Table 5. Response of F9xZ-3 to KLH T Helpers: Dose Response and Kinetics

Dose Response:*

Number of T_H	No Antigen	FL-KLH
0	2,360	3,191
10^3	2,552	2,015
3×10^3	2,008	4,945
10^4	76,381	59,574
3×10^4	295,769	273,072

Kinetics:†

Day	No Helpers	T Helpers
1	6,835	9,026
2	8,456	16,363
3	3,559	26,326
4	8,592	74,669
5	5,673	178,306

*10^3 F9xZ-3 cells were cultured in microtiter wells with 10^5 irradiated spleen cell fillers and the indicated numbers of KLH-enriched helper T cells. 12μg/ml FL-KLH was added to some wells. ^{3}HdT incorporation (cpm) was determined after 5 days of culture.

† 10^3 F9xZ-3 cells were cultured with 10^5 irradiated spleen cell fillers with or without 2×10^4 KLH-enriched helper T cells. On the indicated days, ^{3}HdT uptake was determined.

to this hybridoma, since the proliferation of other B lymphoma lines as well as some T cell lines was modulated by T helper cells. Interestingly, however, this effect seems to be preferential for cells of the lymphoid series, an observation which is being pursued with cloned helper lines and their supernatants via absorption studies.

DISCUSSION

In this presentation, we have described novel approaches for the analysis of B cell triggering and tolerance at the clonal level. These systems, one involving the cloning of hapten-specific B cells and testing of their functional progeny, the other utilizing somatic cell hybridization of hapten-specific cells with B lymphoma lines, have enabled us to begin to analyze B cell heterogeneity and the regulation of B cell growth and differentiation at the clonal level. We have established that different subpopulations of FL-specific CFU-B apparently clone out at different times and under different mitogenic conditions. We have shown that cells grown under certain conditions were responsive to a variety of TI as well as TD antigens; under slightly different conditions (e.g. growth with LPS alone), however, similar numbers of colonies grew but these were of different morphology and were functionally unresponsive *in vitro.* Further evidence indicated that a subpopulation exists with adult B cells grown in SRBC + LPS and which is susceptible to a negative effect caused by FL-SGG. This is a unique population amongst the putative subsets in the adult mouse. While 80% of adult B cells are both IgM and IgD positive, only a portion of these cells are inhibitable in their clonal expansion by anti-δ (12). It will be interesting to determine if there is a correlation between those cells which are sensitive to tolerogen, those cells which are inhibited by anti-δ, and the functional progeny which respond to TD vs TI antigens. Such studies are underway.

An additional level of control relates to the role of macrophages in modulating B cell growth. Although the role of macrophages as accessory cells and in antigen presentation to T cells has long been appreciated, only recently has their participation in supporting B cell growth begun to be recognized. Our experiments indicate that the presence of macrophages during the cloning stage (but not before) (7) modulates the negative effect provided by FL-SGG. However, many of the cells which survive this treatment and grow into discrete colonies are still unresponsive, suggesting a separation in the ability of macrophages to interfere with the tolerance process (cf. ref. 19). Interestingly, we and others have noted (6,11) that CFU-B grown in the presence of macrophages are larger than those colonies grown in the absence of these cells. We are currently testing whether the role of macrophages can be substituted by lymphokines produced by these cells and whether cloned macrophage lines can serve the same purpose.

The second approach which we have used, that of somatic cell hybridization, has provided us with a number of cell lines, one of which can be induced to respond by hapten on LPS to differentiate into antibody secreting cells. This cell line should provide, at the very least, a large number of identical cells for the analysis of the molecular events of one step in B cell activation, that is, terminal differentiation to antibody secretion. Further hybrids are being created in the hope that they will yield clones which express IgM and IgD and represent other stages of the B cell differentiation sequence. Furthermore, these cells may provide a means to look at the events leading to tolerogenesis with large numbers of cloned identical cells.

A by-product of this study was the observation that helper T cells could modulate the growth of this hybrid cell in a manner which did not appear to be antigen-dependent. Preliminary data indicate that a soluble factor is elaborated by such a helper T cell population. Efforts are underway to identify T cell hybridomas which secrete such a factor. A large number of hybridoma clones (derived from these helper T cells used with the R1 thymoma line) are currently being analyzed. It will be important to determine if such growth factors, when obtained from a single clone, are restricted to modulate the growth of B cells alone or whether they may have broader growth promoting activities.

In summary, we have demonstrated that hapten-specific B cell clones can be grown in agar; these specific CFU-B are functionally responsive to a number of different forms of antigen. Such clones can be regulated by mitogenic potentiators, antigen or anti-Ig interactions, macrophages and helper T cells. The study of these CFU-B should now allow the analysis of the functional progeny of a single B cell. The development of hybridoma B cells of a known specificity should further allow an analysis of the molecular biology of B cell triggering and tolerance.

REFERENCES

1. Haas, W. and J. E. Layton, Separation of antigen specific lymphocytes I. Enrichment of antigen-binding cell. J. Exp Med. 141: 1004 (1975).
2. Venkataraman, M., and D. W. Scott, Cellular events in tolerance VII. Decrease in tolerant spleens of PFC precursors stimulated in vitro by specific antigen or mitogen. Cell. Immunol. 47: 323 (1979).
3. Julius, M. and A. Augustin. Helper activity of T cells stimulated in long term culture. Eur. J. Immunol. 9: 671 (1979).
4. Cambier, J. and R. B. Corley. The effect of second signals on the induction of B-cell tolerance: Failure of helper T cells to block tolerance induction. Eur. J. Immunol. 11: 550 (1981).
5. Corradin, G., H. M. Etlinger and J. M. Chiller. Lymphocyte specificity to protein antigens. I. Characterization of the antigen-induced in vitro T cell-dependent proliferative response with lymph node cells from primed mice. J. Immunol. 119: 1048 (1977).
6. Pillai, P. S. and D. W. Scott. Hapten-specific murine colony forming B cells: In vitro response of colonies to fluoresceinated thymus independent antigens. J. Immunol. 126: 1883 (1981a).
7. Pillai, P. S. and D. W. Scott. Hapten-specific murine colony forming B cells. II. Delineation of a tolerance-sensitive subpopulation of colony forming B cells. J. Immunol. 127: XXX (1981b).
8. Metcalf, D., G. J. V. Nossal, N. L. Warner, J. F. A. P. Miller, T. E. Mandel, J. E. Layton and G. A. Gutman. Growth of B lymphocyte colonies in vitro. J. Exp. Med. 142: 1534 (1975).

9. Kincade, P. W., P. Ralph, and M. A. S. Moore. Growth of B lymphocyte clones in semisolid agar culture is mitogen dependent. J. Exp. Med. 143: 1265 (1976).
10. Scott, D. W., M. Venkataraman and J. J. Jandinski. Multiple pathways of B lymphocyte tolerance. Transplant. Rev. 43: 241(1979).
11. Kincade, P. W. Practical aspects of murine B lymphocyte cloning. In: "Methods of Aging Research. Immunology of Aging," Vol. V. CRC Press, West Palm Beach, Fla. (1979).
12. Scott, D. W., P. S. Pillai and S. J. Anderson. Functional hapten-specific B cell clones and hybrids. In: "B Lymphocytes in the Immune Response: Functional, Developmental and Interactive Properties," N. Klinman, D. Mosier, I. Scher and E. Vitetta, Eds., Elsevier North Holland, New York, p. 127 (1981).
13. Gefter, M. L., D. H. Margulies and M. D. Scharff. A simple method for polyethylene glycol-promoted hybridization of mouse myeloma cells. Som. Cell Gen. 3: 231 (1977).
14. Paige, C. J., P. W. Kincade and P. Ralph. Murine B cell leukemia line with inducible surface immunoglobulin expression. J. Immunol. 121: 641 (1978).
15. Slavin, S. and S. Strober. Spontaneous murine B-cell leukemia. Nature 271: 624 (1978).
16. Boyd, A. W., J. W. Goding and J. W. Schrader. The regulation of growth and differentiation of a murine B cell lymphoma. I. Lipopolysaccharide-induced differentiation. J. Immunol. 126: 2461 (1981).
17. Nossal, G. J. V., B. L. Pike and F. L. Battye. Mechanism of clonal abortion tolerogenesis. II. Clonal behavior of immature B cells following exposure to anti-μ chain antibody. Immunology 37: 3030 (1979).
18. Howard, M., D.W. Scott, B. Johnson and W. E. Paul, Long term culture of normal mouse B lymphocytes. In: "B Lymphocytes in the Immune Response: Functional, Developmental and Interactive Properties," N. Klinman, D. Mosier, I. Scher and E. Vitetta, Eds., Elsevier North Holland, New York, p. 141 (1981).
19. Venkataraman, M. and D. W. Scott. Cellular events in tolerance. VIII. Analysis of the proliferative capacity of antigen-binding cells isolated from normal or tolerant spleens. J. Immunol. 124: 607 (1980).

ACKNOWLEDGMENTS

These studies were supported by USPHS grants # AI-10716 (DWS) and AI-16565 (RBC) and Medical Science Training Program grant # GM-07171 (SJA). RBC is a Leukemia Society of America Scholar. We thank Jane Tuttle, Meredith Sedwick and Desireé White for expert technical assistance and Lessie Detwiler for typing the manuscript. We also thank Wendy Havran and Dr. John Cambier for cell sorter analysis.

12

IMMUNOLOGICAL STUDIES WITH TOBACCO MOSAIC VIRUS PROTEIN: DIFFERENTIAL ACTIVATION OF B CELL SUBPOPULATIONS

Donna M. Rennick, Philip R. Morrow and Eli Benjamini

Department of Medical Microbiology, School of Medicine
University of California, Davis

INTRODUCTION

Currently, information concerning the activation requirements of B lymphocytes in different stages of differentiation is limited. It has been known for some time that resting B lymphocytes at different levels of maturation may not respond to the same stimuli in the same way. Activation of immunocompetent lymphocytes from adult animals by T-dependent (TD) antigens can lead to antibody production and memory whereas stimulation of immature lymphocytes (neonatal) may under identical conditions result in tolerance.[1-4] Furthermore, it appears that primary and memory B cells do not exhibit the same requirements for activation under certain experimental conditions. Although some of these results may be explained by quantitative differences, there is evidence for qualitative differences as well.[5-9]

Recently, considerable evidence has accumulated indicating that memory B cells generated in response to T-dependent antigen stimulation consist of distinct subpopulations which can be identified by the sequential appearance or loss of cell surface markers, changing cell size and tissue location.[10-12] It has also been shown that these subpopulations differ in their ability to be activated to clonal expansion and antibody synthesis by T-dependent and T-independent (TI.1 and TI.2) forms of the priming hapten.[13-18] Since these various molecular forms of the hapten differ in their intrinsic mitogenic properties and dependence on accessory cell cooperation in triggering B cells, it is apparent that these memory B cell populations have different molecular and cellular requirements for activation. Presently, there is difficulty in interpreting these findings since the developmental relationship of these memory B cell subsets have not been resolved. Both separate and common lineages have been proposed.[17,18]

Because of the extensive heterogeneity of B lymphocytes, analysis of any one cell lineage or any one state of differentiation is difficult. During the course of our studies, we have developed a system which appears to be well suited to the study of B cell activation at different stages of antigen driven maturation. In these studies tobacco mosaic virus protein (TMVP) is used as the model antigen. Unlike most anti hapten responses, the antibody response to the decapeptide determinant of TMVP (a.a. 103-112) is extremely restricted in heterogeneity. Analysis of these anti decapeptide antibodies by isoelectric focusing patterns and determination of the fine antibody specificities can be employed to indirectly identify the composition of the responding B cells almost at the clonal level. Furthermore, it appears that there is a differential activation of primary and memory B cell subpopulations by various molecular forms of the decapeptide antigen.[9] Given the advantages of this particular system, it was utilized to study the changing activation requirements or triggering thresholds which may occur during the maturation process within a given B cell population.

STUDIES OF B LYMPHOCYTE ACTIVATION BY TMVP

We have previously reported that a.a. 103-112 represent a major antigenic determinant of tobacco mosaic virus (TMVP) in several animal species including several strains of mice.[19] The studies described in this report are concerned with B cell induction by the decapeptide determinant (a.a. 103-112) in CSW mice. We have found that CSW mice immunized with TMVP make a T cell dependent antibody response to decapeptide that is very restricted in heterogeneity. This is based on the finding that individual immune sera separated by isoelectric focusing (IEF) and developed with ^{125}I tyrosylated decapeptide revealed a small number of anti decapeptide spectrotypes (Fig. 1). We have considered the possibility that some of these spectrotypes are due to antibodies with the same V region but with different immunoglobulin heavy chains. This would mean that the number of different antibody specificities present in an individual serum is even more restricted than this study suggests. It is apparent from analyzing the IEF patterns that each serum has several spectrotypes in common with others; however, there are some striking differences in the number and position of other bands. This indicates that although the anti decapeptide antibody response is very restricted, it does not appear to be identical from animal to animal.

In other studies the fine specificities of antidecapeptide antibodies were determined. For this purpose we constructed a

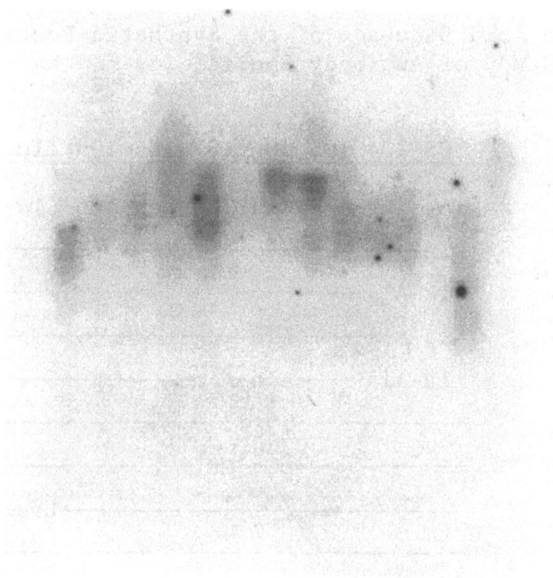

Fig. 1. IEF patterns of sera from TMVP immune CSW mice. The sera were focused in a 5% polyacrylamide gel (pH 4-9 gradient) and developed with ^{125}I-tyr-decapeptide for 12 hrs.

peptide panel (Table 1) consisting of peptides prepared by solid phase peptide synthesis of Merrifield.[20] The amino acid sequence shown at the top of Table 1 is that of decapeptide. The other peptides are closely related derivatives or analogs consisting of single amino acid substitutions. These peptides were radiolabeled and the ability of TMVP immune sera to bind each of them was assessed. It is evident that this panel is capable of distinquishing between individual antidecapeptide antibody species. This is shown by the distinct binding patterns exhibited by monoclonal antibodies obtained from three different hybridoma cell lines of CSW origin (Table 2 - first three columns). The panel has also proven efficient in discerning the fine specificity differences of closely related antibody populations present in individual sera. Approximately 80 sera from TMVP immune CSW mice have been tested against this panel. Nine distinct binding patterns were identified before we started to consistently see repeats of these patterns. Six of the antibody binding patterns are shown in Table 2. It is apparent that the binding patterns obtained with some of these sera can be duplicated by the composite binding of two or all three of the monoclonals. Interestingly, the patterns obtained with other sera are identical to that of an individual monoclonal antibody suggesting that some of these patterns can be duplicated by a single antibody population. However, it is difficult at this time to predict the exact number of different decapeptide specific

Table 1. Amino Acid Sequence of the Synthetic Peptides Used in the Study of Antibody Specificity

PEPTIDE ABBREVIATION	103 104 105 106 107 108 109 110 111 112
Decapeptide*	Thr-Thr-Ala-Glu-Thr-Leu-Asp-Ala-Thr-Arg
Ala-104-112	Ala———————————————————
(Ala)$_2$-105-112	Ala-Ala—————————————————
(Ala)$_4$-107-112	Ala-Ala———Ala—————————————
(Ala)$_5$-108-112	Ala-Ala———Ala-Ala———————————
Tyr$_{106}$	——————Tyr—————————————
Met$_{107}$	————————Met———————————
Leu$_{110}$	———————————————Leu————
Cys$_{112}$	————————————————————Cys

*Decapeptide denotes the amino acid sequence 103-112 of tobacco mosaic virus protein (TMVP).

antibodies which may actually be present in TMVP immune serum. Nevertheless, based on our isoelectric focusing studies, it is known that the number of anti decapeptide antibody populations present in any given serum is very limited indicating that a limited number of different B cell clones are activated. The fact that some sera show distinct antibody specificity differences implies that certain B cell populations are selectively activated during an individual animal's response to immunization with TMVP. This conclusion agrees with the isoelectric focusing data of these sera which demonstrate that they have many anti decapeptide spectrotypes in common with a few unique bands. Depending on the particular combination of B cells induced, a certain isoelectric focusing pattern and fine specificity binding profile with the peptide panel are observed.

In addition to assessing the specificity of the anti decapeptide antibody response, the immunoglobulin isotype distribution was determined by solid phase radioimmunoassay.[22] Briefly, decapeptide was conjugated to a nonrelated carrier, absorbed to flexible polyvinylchloride plates and overlaid with test serum. Following incubation, the plates were washed and then rabbit anti mouse immunoglobulin serum specific for a given immunoglobulin isotype (anti IgG$_1$, IgG$_{2a}$, IgG$_{2b}$, IgG$_3$ and IgM) was added to the appropriate wells. The various test mixtures were then developed with preabsorbed ^{125}I labeled goat anti rabbit immunoglobulin antiserum. The results of these studies have shown that anti decapeptide

Table 2. Direct Binding to Peptides**

PEPTIDE PANEL* (Abbr. designation)	CSW MONOCLONAL ANTIBODIES			INDIVIDUAL CSW SERUM (TMVP IMMUNE)					
	B2	C10	A6	(1)	(2)	(3)	(4)	(5)	(6)
Decapeptide	+++	+++	+++	+++	+++	+++	+++	+++	+++
Ala-104-112	+++	+++	+	++++	+++	+	+++	++	+++
$(Ala)_2$-105-112	++	+++	+	++++	+++	+	-	++	++++
$(Ala)_4$-107-112	+	+++	+	++	++	-	-	++	++
$(Ala)_5$-108-112	-	++	+	++	++	-	-	++	+++
Tyr_{106}	++	+++	++	++	+	+++	-	++	+++
Met_{107}	-	++++	+++	++++	++++	+	++++	++++	+++
Leu_{110}	-	+++	-	++	-	-	-	-	++++
Cys_{112}	++++	-	-	++	++	++++	++++	-	-

*Each peptide was N-acetylated using ^3H acetic anhydride as described by Fearney et al.[21]

**The binding of normal serum with each of the peptides was subtracted from the appropriate determinations as background binding. The net picomoles of each peptide bound by a given serum was divided by the net picomoles of decapeptide bound by that serum. Thus, binding for native decapeptide is expressed as 1.0 and binding with the analogs is expressed relative to this value and adjusted to the following scale: 0.00-0.04 → 0; 0.05-0.09 → ±; 0.10-0.39 → +; 0.40-0.79 → ++; 0.80-1.09 → +++; >1.10 → ++++.

antibodies of each isotype are present with a predominance of IgG_1. Furthermore, individual sera from TMVP immune CSW mice which share the same fine antibody specificity patterns also contain comparable quantities of each immunoglobulin isotype.

The sum of these studies have shown first, there is a very restricted antibody response induced to a single well defined determinant of TMVP. This implies that there is a limited number of inducible B cells with specificity for the decapeptide determinant. Second, analysis of these antibody populations by isoelectric focusing and determination of the fine specificity patterns as well as the immunoglobulin isotype distribution enables us to indirectly identify the responding decapeptide specific B cell clones in a given animal. Third, employing these techniques it should be possible to follow their selective activation or absence of activation under a variety of experimental conditions.

ASSESSMENT OF PRIMARY B CELL ACTIVATION BY DECAPEPTIDE CONJUGATED TO "CLASSICAL" TD AND TI CARRIERS

In view of the previously described studies, we felt that investigation of B cell induction by this protein determinant would give us an advantage in trying to determine the activation requirements of various B cell subpopulations and in establishing their delineation. Our first approach was to immunize several strains of mice with decapeptide conjugated to a variety of chemically dissimilar carriers listed in Table 3. We included several TD carriers and some of the TI carriers recently divided into types TI.1 and TI.2. Some of these carriers were succinylated to permit conjugation of the decapeptide through it's N-terminus only. In all cases, immunization of naive mice with these various conjugates failed to induce a primary anti decapeptide response whereas primary immunization with TMVP induced excellent titers. Furthermore, immunization with each of the decapeptide conjugates failed to induce a primary response when performed under a variety of experimental conditions such as using multiple injections, various dosages, different injection routes, in some cases adjuvants, and conjugates consisting of various decapeptide to carrier ratios.

The apparent nonimmunogenicity of the decapeptide antigens prompted us to investigate several properties of these conjugates. These studies demonstrated that the decapeptide is presented on each of the conjugates in an antigenic form since they bind with anti decapeptide antibodies elicited by TMVP immunization and they can inhibit the binding between TMVP immune serum and ^{125}I tyr-decapeptide. It was established that immunization with these conjugates does not suppress an anti decapeptide response to TMVP when injected prior to or at the same time as TMVP challenge. It

Table 3. Carriers Used in the Preparation of Decapeptide Conjugates

"TD" Carriers	"TI" Carriers
Human gamma globulin (HGG)	Brucella abortus (BA)
Succinylated human gamma globulin (SHGG)	Succinylated ficoll (S-ficoll)
Succinylated lysozyme	Aminoethylcarboxymethyl ficoll (AECM-ficoll)
Poly-(L)-Lysine	Lipopolysaccharide (LPS)

was also shown that immunization with any of the conjugates prior to TMVP did not enhance the subsequent anti decapeptide response as compared to animals immunized with only TMVP. Therefore, we concluded that these various forms of the decapeptide antigen do not appear to "silently" prime naive B cells in the absence of a detectable antibody response. Those conjugates consisting of protein carriers were capable of inducing carrier specific T helper cells. Finally, primary antibody responses could be elicited to other epitopes similarly conjugated to these carriers. Currently, the inability of these decapeptide antigens to induce naive B cells specific for decapeptide is still under investigation. We feel that such studies are warranted in view of the possibility that these results may further delineate certain regulatory signals required to initiate primary B cell activation.

ASSESSMENT OF MEMORY B CELL ACTIVATION BY VARIOUS DECAPEPTIDE CONJUGATES

Since we could not demonstrate any inherent properties of the decapeptide conjugates which should prevent them from functioning as immunogens, we proceeded to assess their ability to activate memory B cells using an adoptive transfer system. Spleen cells from mice immunized with TMVP (source of decapeptide specific memory B cells) were transferred into irradiated syngeneic recipients. Following transfer, naive and carrier primed recipients were secondarily challenged with TMVP or one of the decapeptide conjugates. Two weeks later their sera were assayed for antibodies capable of binding ^{3}H-N-acetyl-decapeptide. The results of such an experiment are shown in Fig. 2. As expected, secondary challenge with TMVP elicited a high anti decapeptide titer. In contrast to the inability of decapeptide-SHGG to activate a primary response, secondary challenge of memory B cells with this conjugate induced an anti decapeptide response comparable to that of TMVP. However, an antibody response was elicited in only those mice which had

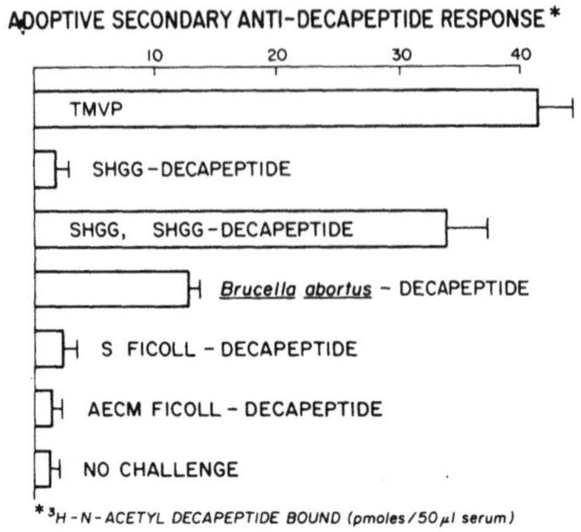

Figure 2

been primed to the SHGG carrier in complete Freund's adjuvant. Since the requirement for carrier priming is due to a dependence on theta positive cells (data not shown), this conjugate represents a TD form of the decapeptide antigen. The decapeptide-BA conjugate, which was also shown to be nonimmunogenic in naive mice, stimulated a significant but consistently lower secondary antibody response than the two TD forms of the antigen (TMVP and decapeptide-SHGG). We have found that the ability of decapeptide-BA to simulate decapeptide memory B cells has little or no dependence on the presence of theta positive cells (data not shown). Surprisingly, neither decapeptide-succinylated ficoll nor decapeptide-AECM ficoll activated TMVP primed B cells.

Although decapeptide-SHGG and decapeptide-BA are incapable of stimulating primary responses, these results show that they are able to stimulate antibody production in memory B cells. This differential activation of primary and memory B cells is in agreement with other investigations which indicate qualitative differences in the activation requirements between these two maturational stages.[5-9] Furthermore, the ability of decapeptide-BA to activate TMVP primed cells is consistent with the findings in other systems which have shown that memory B cells primed by the TD form of a hapten can be cross stimulated by the same hapten present on the TI.1 type carrier Brucella abortus.[13,18] Decapeptide-S ficoll and decapeptide-AECM ficoll represent TI.2 forms of our protein determinant. Their inability to activate decapeptide specific memory B cells, in contrast to secondary stimulation obtained with decapeptide-BA, supports the concept held by some that haptenated TI.1 and TI.2 antigens may be recognized by separate subsets of B cells[18,23,24] even at the memory B cell level.[18] One possible reason for the lack of responsiveness to decapeptide-ficoll antigens in our system is that the induction of a clonally restricted B cell response by decapeptide within the context of TMVP does not include certain TI.2 responsive B cell subsets. This possibility is currently being studied.

The reasons for the differential activation of primary and memory B cells by decapeptide-SHGG and decapeptide-BA are not yet clear. Nevertheless, it does afford us the opportunity to use these antigens to ask specific questions concerning the activation of memory B cell subsets without interference from the induction of a primary response at the same time. Therefore, we focused on the observation that memory B cells induced by a strict T-dependent antigen, TMVP, could be activated by decapeptide-SHGG and decapeptide-BA via mechanisms which differ markedly in their dependence on accessory T cell help. The obvious question to ask was whether the TD and TI.1 forms of the decapeptide antigen were stimulating the same population of memory cells? Alternatively, does TMVP prime a heterogeneous mixture of B cells originating from distinct lineages whose requirements for activation are different? Experiments designed to answer these questions are discussed in the following section.

DO DECAPEPTIDE-SHGG AND DECAPEPTIDE-BA STIMULATE THE SAME POPULATION OF DECAPEPTIDE SPECIFIC MEMORY B CELLS OR DISTINCT SUBSETS?

We used several approaches to determine the relationship between those memory B cells which are activated by the TD decapeptide antigens (TMVP and decapeptide-SHGG) and those which respond to the TI decapeptide antigen (decapeptide-BA). Our first approach was to determine if the injection of two decapeptide antigens simultaneously would result in an additive antibody response. An additive response would indicate that each antigen stimulated separate B cell subsets. As shown in Fig. 3, the antibody titer elicited by two decapeptide antigens given together was never more than the higher response to one of the antigens alone, thus, overlapping subsets of memory B cells are being activated by these various antigens.

To further corroborate this preliminary conclusion, we attempted to distinguish between the memory B cells responding to secondary stimulation with TMVP, decapeptide-SHGG or decapeptide-BA by analyzing the anti decapeptide antibody response. To this end, mice were transferred with the same pool of TMVP immune spleen cells from donors exhibiting a given anti decapeptide antibody specificity profile. The results shown in Table 4 indicate that each immunogen elicited the same antibody populations since the composite binding gave identical fine specificity profiles. These findings were confirmed by the isoelectric focusing patterns obtained with these sera. Furthermore, assessment of the immunoglobulin isotype composition of decapeptide specific

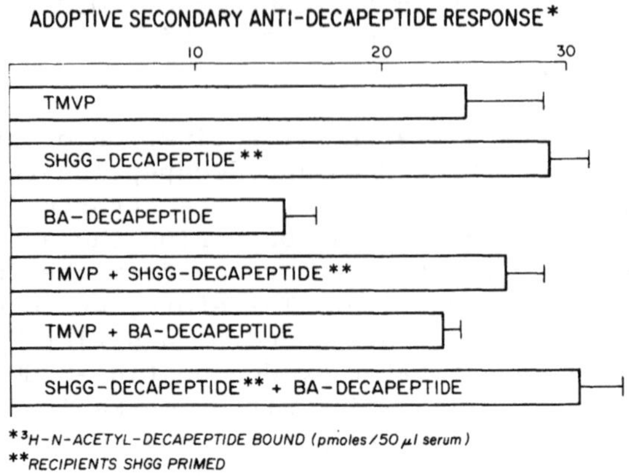

Figure 3

antibodies in these same sera revealed that the isotype composition induced by decapeptide-SHGG and decapeptide-BA was identical to that induced by TMVP (Fig. 4). Therefore, it appears that the TD and TI.1 decapeptide antigens do not preferentially activate TMVP memory B cells of a given specificity or with the potential to

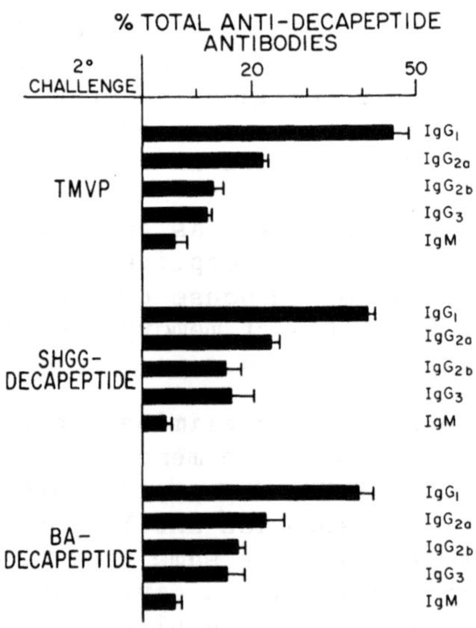

Fig. 4. The relative decapeptide specific Ig isotype composition following secondary challenge with different immunogens. The composition is expressed as a ratio of decapeptide binding by Ig of a given isotype (per 20 µl serum) to the total decapeptide binding Igs (per 20 µl serum).

Table 4. Binding of Antibodies with Peptides

PEPTIDE PANEL	CSW MONOCLONAL ANTIBODIES			CSW ADOPTIVE 2° CHALLENGE WITH:		
(Abbr. designation)	B2	C10	A6*	BA-DECAPEPTIDE	SHGG-DECAPEPTIDE	TMVP**
Decapeptide	+++	+++	+++	+++	+++	+++
Ala-104-112	+++	+++	+	++	++	++
$(Ala)_2$-105-112	++	+++	+	++	++	++
$(Ala)_4$-107-112	+	+++	+	++	++	++
$(Ala)_5$-108-112	-	++	+	++	++	++
Tyr_{106}	++	+++	++	+	+	+
Met_{107}	-	++++	+++	++++	++++	++++
Leu_{110}	-	+++	-	-	-	-
Cys_{112}	++++	-	-	-	-	±

*CSW monoclonal anti decapeptide antibodies are included in each determination as indicators of specificity. The binding values are calculated as described in the legend of Figure 3.

**The sera from mice secondarily challenged with BA-decapeptide SHGG-decapeptide or TMVP were individually tested for binding to each peptide. The peptide binding values reported are an average of determinations of six individual sera.

synthesize a certain immunoglobulin isotype. In fact, there is no "carrier" effect on the isotype composition of a secondary anti decapeptide response since it is unaltered from that originally established by the priming antigen, TMVP. Based on the combined results of these studies, we have concluded that the TD and TI.1 decapeptide antigens stimulate overlapping populations of memory B cells.

ASSESSMENT OF NEW MEMORY INDUCTION BY DECAPEPTIDE-SHGG AND DECAPEPTIDE-BA

Although these studies show that certain decapeptide conjugates are capable of inducing memory B cells to differentiate into antibody forming cells, we felt it was important to test their ability to induce proliferation leading to memory regeneration. This seemed especially pertinent in view of the inability of these conjugates to activate primary B cells. To test for memory regeneration, TMVP primed spleen cells were transferred into naive and carrier primed mice. Following transfer, the recipients were challenged with TMVP or one of the decapeptide conjugates and their antibody titers assessed. The mice were rested for eight weeks and then rechallenged according to the scheme in Fig. 5. The detection of an enhanced antibody response following two challenges was interpreted as evidence for an expanded memory B cell population as a result of the first antigen stimulation.

The results shown in Fig. 5 suggest several important points. It is apparent that two challenges with decapeptide-SHGG gave an

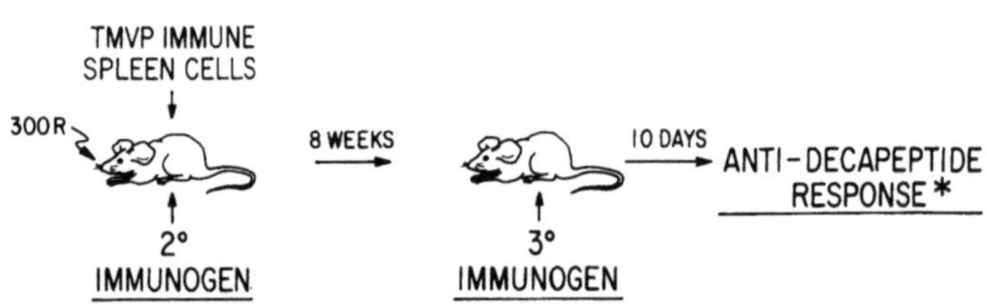

2° IMMUNOGEN		3° IMMUNOGEN	ANTI-DECAPEPTIDE RESPONSE *
NO CHALLENGE	→	SHGG-DECAPEPTIDE	32 ± 1.7
SHGG-DECAPEPTIDE	→	SHGG-DECAPEPTIDE	81 ± 4.2
BA-DECAPEPTIDE	→	BA-DECAPEPTIDE	2 ± 0.7
BA-DECAPEPTIDE	→	SHGG-DECAPEPTIDE	21 ± 0.5

*pmoles ^{3}H-N-ACETYL-DECAPEPTIDE BOUND/50 μl SERUM

Figure 5

enhanced titer (line 2) as compared to that group receiving one decapeptide-SHGG injection (line 1). Therefore, secondary challenge with this antigen not only induces antibody synthesis but leads to memory regeneration as well. This just gives emphasis to the finding that the activation requirements of memory B cells leading to differentiation and proliferation may be qualitatively different from those of primary B cells since primary B cells are nonresponsive to this antigen. The results also show that two injections with decapeptide-BA does not lead to an enhanced antibody response, in fact, it fails to elicit a significant response at all (line 3). There are several explanations for this observation. It is possible that the secondary challenge with decapeptide-BA, although it induced an antibody response initially (Fig. 3), may lead eventually to blocking or a state of specific B cell tolerance so that the memory B cells become nonresponsive to further stimulation. This possibility does not seem likely in view of our earlier findings which have shown that naive mice immunized with this conjugate prior to or in combination with TMVP respond with the expected anti decapeptide titers. The second possibility is that secondary stimulation with decapeptide-BA differentiates memory B cells to antibody forming cells without memory regeneration and this could lead to a "depletion" of decapeptide specific memory B cells. If this is the case, what happens if these mice are rechallenged with decapeptide-SHGG? As the data show (Fig. 5 - line 4) a significant but reduced titer was obtained. This implies that decapeptide-BA is capable of activating only a portion of the memory B cells transferred. Those which remain following this challenge can be activated by the T-dependent antigen, decapeptide-SHGG. We obtained the same results when we followed decapeptide-BA challenge with TMVP (data not shown).

In sum, decapeptide-SHGG induces proliferation leading to memory regeneration, but decapeptide-BA does not. Further, there appears to be two functionally distinct subsets of memory B cells generated in response to TMVP immunization. Those which require TD antigen stimulation (designated M_1) and those which can be activated by TD and TI.1 decapeptide antigens (designated M_2). The results of this experiment appear to contradict the previously described studies which indicated that TMVP, decapeptide-SHGG and decapeptide-BA were activating the same memory B cell populations. This was based on analysis of the secondary antibody response induced by each antigen by isoelectric focusing, determination of the fine antibody specificity profiles and isotype compositions. It was further supported by the finding that secondary stimulation with a combination of the TD and TI antigens never resulted in an additive antibody response. These previous results and the results obtained in the experiment shown in Fig. 5 can be reconciled by hypothesizing that the two memory B cells subsets, M_1 and M_2, represent two stages of maturation sharing a common lineage but with different requirements for activation.

ASSESSMENT OF THE DEVELOPMENTAL RELATIONSHIP OF MEMORY B CELL SUBSETS RESPONSIVE TO TI AND TD FORMS OF THE DECAPEPTIDE ANTIGEN

To test for a developmental relationship between decapeptide specific memory subsets M_1 and M_2, the experiment depicted in Fig. 6 was performed. TMVP immune spleen cells were transferred into irradiated recipients along with SHGG carrier primed cells. All of the groups were secondarily challenged with decapeptide-BA. After six weeks of rest, these mice were given a tertiary challenge with one of the decapeptide antigens as indicated. Since some of these antigens elicit antibodies which persist for a long period of time, the recipients were rested for six weeks and then the lymph node cells and spleen cells were transferred into new recipients. Following the second transfer step, each group was challenged with antigen according to the scheme in Fig. 6.

The purpose of challenging all of the mice with decapeptide-BA was to "deplete" the transferred TMVP memory B cells of all decapeptide-BA responsive cells. The first two groups (line 1 and 2) are the controls for this step. It is apparent that secondary challenge with decapeptide-BA renders the recipients nonresponsive to further stimulation with this conjugate. One of the test groups was given a tertiary injection with decapeptide-SHGG in the presence of carrier primed cells (line 3). Following the second transfer step, they were rechallenged with decapeptide-BA. It is evident that challenge with decapeptide-SHGG regenerated new memory B cells capable of responding to a quaternary challenge with decapeptide-BA. New decapeptide-BA responsive B cells were also regenerated following a tertiary challenge with TMVP (line 5). The fact that

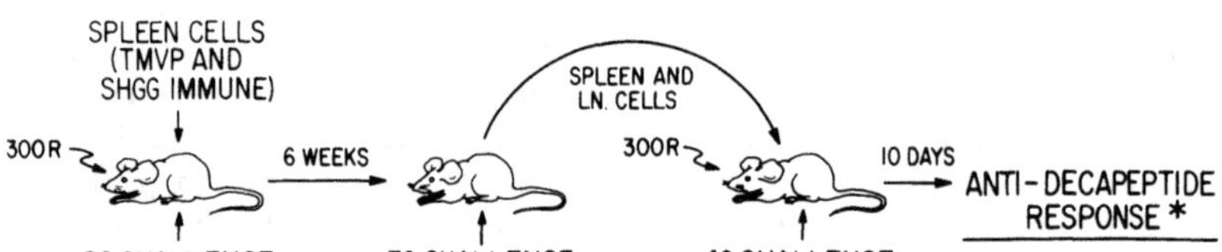

2° CHALLENGE	3° CHALLENGE	4° CHALLENGE	ANTI-DECAPEPTIDE RESPONSE *
	—	BA-DECAPEPTIDE	<1
	BA-DECAPEPTIDE	BA-DECAPEPTIDE	<1
BA-DECAPEPTIDE	SHGG, SHGG-DECAPEPTIDE	BA-DECAPEPTIDE	11 ± 0.7
	SHGG-DECAPEPTIDE	BA-DECAPEPTIDE	<1
	TMVP	BA-DECAPEPTIDE	19 ± 0.9
	SHGG, SHGG-DECAPEPTIDE	SHGG-DECAPEPTIDE	29 ± 1.1

*pmoles ^{3}H-N-ACETYL-DECAPEPTIDE BOUND/50 μl SERUM

Figure 6

the regeneration step is T cell dependent is demonstrated by the finding that mice challenged with decapeptide-SHGG in the absence of carrier primed cells remained nonresponsive to a subsequent injection with decapeptide-BA (line 4).

In those mice receiving a tertiary challenge with TMVP, it is possible that new decapeptide-BA responsive memory cells were recruited from primary B cells. However, this is not the case with those animals receiving a tertiary injection with decapeptide-SHGG since this conjugate is incapable of activating primary decapeptide specific B cells. Therefore, those memory B cells still requiring T cell dependent antigen stimulation (M_1) contain the direct precursors to more mature memory B cells (M_2) which are stimulated by the TI.1 form of the antigen to differentiate into antibody forming cells. This developmental relationship demonstrates that there are two distinct stages of memory B cell maturation with different requirements for T-dependent antigen stimulation.

CONCLUSION

The sum of these studies have lead to the preliminary conclusions depicted in Fig. 7. Three functionally distinct populations of B cells have been identified (primary and memory B cell subsets M_1 and M_2). Several lines of evidence have shown that these three B cell populations share common lineages. It has been shown that they represent different stages of maturation by demonstrating their direct precursor-successor relationships. Finally, B cells at different stages of differentiation do not have the same requirements for activation based on their ability or inability to respond to the various TD and TI forms of the decapeptide antigen.

We are currently pursuing this line of investigation since it seems reasonable that those signals leading to the production of immunocompetent B cells would be different from those required to

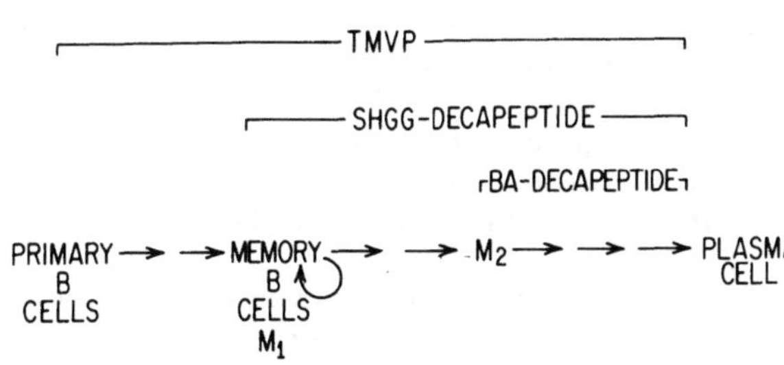

Fig. 7. Distinct stages of B lymphocyte maturation identified by their differential activation by various molecular forms of the decapeptide determinant.

initiate steps leading to terminal differentiation. Such differences would allow for a more efficient regulation of each stage of maturation and may even provide a possible fail-safe mechanism against an inadvertant activation and rapid depletion of large numbers of immunocompetent cells by terminal differentiation.

ACKNOWLEDGEMENTS

The authors wish to acknowledge the excellent technical assistance of Mrs. Cherry Y. Leung, Samuel A. Lewis and Jeffrey S. Krebs.

This research was supported in part by National Science Foundation Grants PCM 76-81361, PCM 77-15927 and PCM 81-03264; Grant AI-05966 from the U.S. Public Health Service; and Grant IM-291 from the American Cancer Society.

REFERENCES

1. E. S. Metcalf and N. R. Klinman, In vitro tolerance induction of neonatal murine B cells, J. Exp. Med. 143:1327 (1977).
2. J. C. Cambier, E. S. Vitetta, J. W. Uhr and J. R. Kettman, B cell tolerance. II. Trinitrophenyl human gammaglobulin-induced tolerance in adult and neonatal murine B cells responsive to thymus-dependent and independent forms of the same hapten, J. Exp. Med. 145:778 (1977).
3. H. M. Etlinger and J. M. Chiller, Maturation of the lymphoid system. II. Characterization of the cellular levels of unresponsiveness induced in neonates by a T-dependent antigen that is an obligate immunogen in adults, J. Immunol. 122:2564 (1979).
4. G. J. V. Nossal and B. L. Pike, Evidence for the clonal abortion theory of B lymphocyte tolerance, J. Exp. Med. 141:904 (1975).
5. N. R. Klinman, J. L. Press and G. P. Segal, Overlap stimulation of primary and secondary B cells by cross-reactive determinants, J. Exp. Med. 138:1276 (1973).
6. D. H. Katz, The allogeneic effect, in: "Immunopotentiation," Ciba Foundation Symposium No. 18. G. E. W. Wolstenholme and J. Knight, eds., Associated Scientific Publishers, Amsterdam (1973).
7. S. Cammisuli and C. Henry, Role of membrane receptors in the induction of an in vitro secondary anti-hapten response. II. Antigen-immunoglobulin receptor interaction is not required for B memory cell proliferation, Eur. J. Immunol. 8:622 (1978).
8. S. Cammisuli, C. Henry and L. Wofsy, Role of membrane receptors

in the induction of an in vitro secondary anti-hapten response. I. Differentiation of B memory cells to plasma cells is independent of antigen-immunoglobulin receptor interaction, Eur. J. Immunol. 8:656 (1978).

9. D. M. Rennick, P. R. Morrow, C. Y. Leung, R. J. Scibienski and E. Benjamini, Differential activation of primary and memory B. lymphocytes, J. Immunol. 125:1262 (1980).
10. S. Strober, Immune function cell surface characteristics and maturation of B cell subpopulations, Transplant. Rev. 24: 84 (1975).
11. M. V. Hobbs and T. L. Feldbush, Antigen modulation of the immune response. VI. Rate of large memory cell appearance in lymph nodes and thoracic duct lymph, Cell. Immunol. 50:30 (1980).
12. H. G. Durkin and G. J. Thorbecke, Homing of immunologically committed lymph node cells to germinal centers in rabbits, Nature-New Biol. 238:53 (1972).
13. S. P. Lerman, T. J. Romano, J. J. Mond, M. Heidelberger and G. J. Thorbecke, Induction of primary and inhibition of secondary antibody response to hapten by hapten conjugates of type III pneumococcal polysaccharide, Cell. Immunol. 15: 321 (1975).
14. W. E. Paul, B. Benacerraf, B. W. Siskind, E. A. Goidl and R. A. Reisfeld, The anamnestic antibody response to type III specific pneumococcal polysaccharide, J. Exp. Med. 130:77 (1969).
15. H. Braley-Mullen, Secondary IgG responses to type 3 pneumococcal polysaccharide. III. T cell requirement for development of B memory cells, Eur. J. Immunol. 7:775 (1977).
16. T. V. Tittle and M. B. Rittenberg, Expression of IgG memory response in vitro to thymus-dependent and thymus-independent antigens, Cell. Immunol. 35:180 (1978).
17. D. T. Umetsu, J. M. Chapman-Alexander and G. J. Thorbecke, Cross-priming of murine B cells with TNP conjugates of hemocyanin and ficoll: characteristics of primed B cells responding to both antigens. J. Immunol. 123:396 (1979).
18. T. V. Tittle and M. B. Rittenberg, Distinct subpopulations of IgG memory B cells respond to different molecular forms of the same hapten, J. Immunol. 121:936 (1978).
19. E. Benjamini, Immunochemistry of the tobacco mosaic virus protein, in: "Immunochemistry of Proteins," M. Z. Atassi, ed., Plenum Press, New York (1977).
20. R. B. Merrifield, Solid phase peptide synthesis. III. An improved synthesis of bradykinin, Biochem. 3:1385 (1964).
21. F. J. Fearney, C. Y. Leung, J. D. Young, and E. Benjamini, The specificity of antibodies to a peptide determinant of the tobacco mosaic virus protein induced by immunization with the peptide conjugate, Biochem. Biophys. Acta 243:509 (1971).
22. T. T. Tsu and L. A. Herzenberg, Solid-phase radioimmune assays,

in: "Selected Methods in Cellular Immunology," B. B. Mishell and S. M. Shiigi, eds., W. H. Freeman and Co., San Francisco (1980).

23. G. K. Lewis and J. W. Goodman, Carrier-directed anti-hapten responses by B-cell subsets, J. Exp. Med. 146:1 (1977).

24. H. S. Boswell, A Ahmed, I. Scher and A. Singer, Role of assessory cells in B cell activation. II. The interaction of B cells with accessory cells results in the exclusive activation of an Lyb^{5+} B cell population, J. Immunol. 125: 1340 (1980).

13

ANTIGEN BRIDGING IN THE INTERACTION OF T HELPER CELLS AND B CELLS*

Joel W. Goodman, Danute E. Nitecki, Sherman Fong and Zehra Kaymakcalan

Department of Microbiology and Immunology
University of California
San Francisco, California 94143

Antigen-bridging models of the interaction between T and B lymphocytes rose to prominence in the late 1960s, based principally on two different experimental approaches (1-3). One involved the "carrier effect", in which physical association between hapten and carrier is required for a secondary anti-hapten antibody response in animals primed separately with hapten and carrier, and the other involved thymus-marrow reconstitution of irradiated animals, demonstrating that hapten and carrier were recognized by different cell types in the anti-hapten antibody response.

The approach we have taken to the general question of mechanisms of cell cooperation has made use of small, structurally defined antigens in which the hapten and carrier components have spatial relationships which are reasonably well understood (4). These antigens are epitomized by L-tyrosine-p-azobenzenearsonate (RAT), which induces cellular immunity without antibody formation in several animal species and serves as a carrier for haptens (5,6). Early evidence in this system favoring bridging models stemmed from enhanced anti-hapten antibody responses in guinea pigs immunized with bifunctional antigens in which the hapten and carrier epitopes were separated by a spacer (DNP-spacer-RAT) as compared to when they were not (DNP-RAT) (7), and the requirement for rigid spacers separating the epitopes of symmetrical bifunctional antigens (RAT-spacer-RAT) in cases where the epitopes tend to associate intramolecularly due to charge complementarity (8,9).

*Supported by NIH Grant AI 05664.

The ability to construct synthetic bifunctional antigens with rigid spacers (DNP)-(proline)$_n$-RAT) permitted more rigorous probing of the validity of antigen bridging. Consider the plausibility that T-dependent B cell activation requires contact between the B cell and either a helper T (T_H) cell itself or a soluble product released by the T_H cell, with antigen serving as the focusing agent in either case. For this argument, we shall ignore the role of the macrophage, which has definitively been shown to be essential for activation of the T_H cell, but for whose role in interaction between B cells and activated T_H cells there is less evidence. The argument is supported by abundant evidence favoring genetic restriction of T-B collaboration in at least some circumstances (10,11), and by the elaboration of antigen-specific helper factors bearing markers coded by the major histocompatibility complex (12). It is independent of the possible existence of more than one type of T_H cell (13). If the T cell or its product recognizes the carrier epitope of a bifunctional antigen which is bound to a B cell by its haptenic epitope, and if contact between the B cell and a "cell interaction structure" on the T cell or its product is required for B cell activation, then it should be possible to construct bifunctional antigens with rigid spacers of dimension sufficient to prohibit cell contact. Rigid spacers are required for this purpose to prevent flexing of the spacer chain with consequent indeterminate variance of the distance between the carrier and haptenic epitopes. If such proved to be the case, the maximum permissible distance between the epitopes which still allowed B cell activation could be estimated.

Bifunctional antigens with rigid polyprolyl spacers (DNP-(PRO)$_n$-RAT), were synthesized by the solid phase technique (14). The spacers comprised as many as 40 proline residues. Guinea pig were immunized in the footpads with varying molar quantities of the antigens emulsified in complete Freund's adjuvant. The antigens used and the anti-DNP antibody responses mounted by the animals are summarized in Table I. The PRO_{10} and PRO_{22} compounds evoked responses, measured as precipitating antibody, that peaked about 3 weeks after immunization at levels of 150 to 200 µg of anti-DNP antibody per milliliter of serum. These responses were quantitatively and kinetically similar to those obtained using bifunctional antigens with flexible spacers (7). The optimal antigen dose of those tested was 1.0 µmol in each instance. This dose of DNP-PRO-RAT did not raise significant anti-DNP responses, suggesting that a single proline residue is an ineffective spacer for implementing cell interaction. Of particular interest, DNP-(PRO)$_{31}$-RAT was negative at all doses and times assayed. Extending the spacer to 40 proline residues also gave negative results, indicating that the absence of a response to DNP-(PRO)$_{31}$-RAT was not idiosyncratic.

TABLE I

Antibody Responses to Bifunctional Antigens with Polyprolyl Spacers

Antigen	Spacer Length[a]	Anti-DNP Antibody
1.0 μM	Å	μg/ml
DNP-PRO-RAT	3	0
DNP-$(PRO)_{10}$-RAT	31	70 ± 26[b]
DNP-$(PRO)_{22}$-RAT	69	143 ± 14
DNP-$(PRO)_{31}$-RAT	97	0
DNP-$(PRO)_{40}$-RAT	125	0
DNP-$(PRO)_{16}$SAC-$(PRO)_{15}$-RAT	-	153 ± 26[c]

[a]Based on axial translation of 3.12 Å per proline residue in poly-L-proline (15,16).

[b]Arithmetic mean ± S.E.; 3-6 guinea pigs per group.

[c]Compound tested in separate experiment with DNP-$(PRO)_{22}$-RAT and DNP-$(PRO)_{31}$-RAT. Value normalized to response against DNP-$(PRO)_{22}$-RAT.

Assessment of the distance between the two epitopes in these antigens relies on several assumptions concerning the rigidity and dimension of polyproline chains, which can exist in two alternative forms (15,16). Form I is a cis right-handed helix with an axial translation of 1.9Å per residue. This comformation is preferentially assumed in non-polar solvents. Form II i a trans left-handed helix with three residues per helical turn and an axial translation of 3.12Å per residue. The trans conformation is assumed in aqueous (polar) solvents. The form II coordinates were used in calculating the distances between carrier and haptenic epitopes of the bifunctional antigens (Table I).

The reliability of the figures is supported by several studies. Schimmel and Flory (17) showed that the mean square end-to-end distance of a hexadecamer of L-proline closely approximates the value obtained from the helix coordinates for form II. For longer chain lengths of about 50 residues, intrinsic viscosities are in excellent agreement with predicted values (18). Moreover, in studies more relevant to the work presented here, in which oligomers of proline were used to separate donor and acceptor chromophores in energy transfer measurements, the observed efficiency of energy transfer as a function of distance was also in excellent agreement with predictions (19). These observations, in aggregate, provide solid support for the assumption that polyproline chains with 10-50 residues behave as rigidly extended rods in solution (and, presumably, at the locus of cell interaction, although this clearly represents a greater extrapolation).

The figures thus derived lead to the conclusion that epitopes separated by 69A (PRO_{22}) are fully operative for cell cooperation, whereas a span of 97A (PRO_{31}) obviates delivery of the helper signal to the B cell. The maximum permissible separation could, of course, be more closely estimated by fabricating antigens with intermediate spacers, but the principle has been established and the additional synthetic work required would not justify the information to be gained. It might be argued that the long polyprolyl spacer antigens which do not induce anti-DNP responses induce suppression instead. This possibility is unlikely inasmuch as all animals, regardless of the antigen they received, developed comparable arsonate-specific delayed skin reactivity when tested with ABA-bovine serum albumin conjugates (14). Some animals developed small quantities of anti-polyprolyl antibody, but there were no significant differences between the groups. Thus, animals immunized with 1.0 μmol of the $(PRO)_{10}$, $(PRO)_{22}$ and $(PRO)_{31}$ compounds made 23±23, 31±23 and 16±14 μg of anti-polyproline antibody per milliliter of serum, repectively (14).

The model can be tested further by constructing an antigen with a non-permissive polyproline spacer in which a single proline residue in the middle of the chain is replaced by a different group, permitting a bend or kink in the spacer. If the model is correct, such an antigen might be expected to induce anti-DNP antibody responses. Accordingly, the bifunctional antigen DNP-$(PRO)_{16}$-SAC-$(PRO)_{15}$-RAT was synthesized with a 6-aminocaproyl (SAC) group in the middle of the $(PRO)_{31}$ spacer chain. Guinea pigs immunized with this compound responded with anti-DNP antibody levels fully comparable to those mounted against the bifunctionals with $(PRO)_{10}$ and $(PRO)_{22}$ spacers (Table I). Thus, the introduction of a flexible residue in the middle of a non-permissive polyproline spacer restored the antibody response. This result is very difficult to explain without invoking some type of antigen bridging mechanism in B cell activation.

The present findings are clearly consistent with both the carrier effect (2,3) and the MHC restriction of T helper cell interaction with B cells (10,11), alluded to earlier. These phenomena are most readily explicable by postulating the focusing of T help on B cells by antigen. Whether T help can recognize unmodified antigen sufficiently well to be focused in this way, as indicated by some reports (12), is still problematic, but the responses to rigidly spaced bifunctional antigens in the present study are difficult to reconcile with a single T cell receptor site which recognizes a complex of nominal antigen and a cell surface structure ("altered self"). They are compatible with a model in which functional T help bears a receptor for nominal antigen and a cell interaction structure coded by the MHC (12), through which the triggering signal is delivered to the B cell.

A scenario that satisfies most of the observations about T_H cells is one in which the receptor on the surface of the cell and the soluble helper factor secreted by the cell do not have identical reactivities. Thus, helper T cells do not "see" nominal antigen sufficiently well to be enriched by affinity chromatography, "suicided" or activated. They are activated in an MHC-restricted fashion by antigen associated with antigen-presenting cells. Once activated, they release soluble antigen-specific helper factors which find antigen-binding B cells by virtue of their ability to recognize nominal carrier epitopes. Only one helper factor has been studied in any detail thus far (12). This "TGAL"-specific factor binds nominal antigen and bears H-2-coded determinants, features which are required to explain antigen bridging and MHC-restriction in T-B cell interaction. It remains to be seen if additional evidence supports this model.

REFERENCES

1. P.A. Bretscher and M. Cohn. Minimal model for the mechanism of antibody induction and paralysis by antigen. Nature 220: 444 (1968).
2. N.A. Mitchison, K. Rajewsky, and R.B. Taylor. Cooperation of antigenic determinants and of cells in the induction of antibodies, in: "Prague Sym. on Developmental Aspects Formation and Structure." J. Sterzl and T. Riha, eds., Academic Press, New York (1970).
3. K. Rajewsky, V. Schirrmacher, S. Nase, and N.K. Jerne. The requirement of more than one antigenic determinant for immunogenicity. J. Exp. Med. 129:1131 (1969).

4. J.W. Goodman. Antigenic determinants and antibody combining sites, in: The Antigens, Vol. III, M. Sela, ed., Academic Press, New York (1975).
5. S.S. Alkan, D.E. Nitecki, and J.W. Goodman. Antigen recognition the immune response. The capacity of L-tyosine-p-azobenzenearsonate to serve as a carrier for a macromolecular hapten. J. Immunol. 107:353 (1971).
6. J.W. Goodman, S. Fong, G.K. Lewis, R. Kamin, D.E. Nitecki, and G. Derbalian. Antigen structure and lymphocyte activation. Immunol. Rev. 39:36 (1978).
7. S.S. Alkan, E.B. Williams, D.E. Nitecki, and J.W. Goodman. Antigen recognition and the immune response. Humoral and cellular responses small monofunctional and bifunctional antigen molecules. J. Exp. Med. 135:1228 (1972).
8. M.E. Bush, S.S. Alkan, D.E. Nitecki, and J.W. Goodman. Antigen recognition and the immune response. Self-help with symmetrical bifunctional antigen molecules. J. Exp. Med. 136:1478 (1972).
9. J.W. Goodman, C.J. Bellone, D. Hanes, and D.E. Nitecki. Antigen structural requirements for lymphocyte triggering and cell cooperation, in: Progress in Immunology II, Vol. 2, L. Brent and J. Holbrow, eds. North Holland Publishing Co., (1974).
10. J. Andersson, M.H. Schreier, and F. Melchers. T-cell dependent B-cell stimulation is H-2 restricted and antigen dependent only at the resting B cell level. Proc. Nat. Acad. Sci. USA 77:1612 (1980).
11. B. Jones and C.A. Janeway, Jr. Cooperative interaction of B lymphocytes with antigen-specific helper T lymphocytes is MHC restricted. Nature 292:547 (1981).
12. R.N. Apte, I. Lowy, R. De Baetselier, and E. Mozes. Establishment and characterization of continuous helper T cell lines specific to poly(L Tyr, L Glu)-poly(DL Ala)-poly(L Lys). J. Immunol. 127:25 (1981).
13. C.A. Janeway, Jr., R.A. Murgita, F.I. Weinbaum, R. Asofsky, and H.W. Wigzell. Evidence for an immunoglobulin-dependent antigen-specific helper T cell. Proc. Nat. Acad. Sci. USA 74:4582 (1977).
14. S. Fong, D.E. Nitecki, R.M. Cook, and J.W. Goodman. Spatial requirements between haptenic and carrier determinants for T-dependent antibody responses. J. Exp. Med. 148:817 (1978).
15. V.D. Gupta, R.D. Singh, A.M. Dwivedi. Vibrational spectra and dispersion curves of poly-L-proline II chain. Biopolymers 12:1377 (1973).
16. P.M. Cowan and S. McGavin. Structure of poly-L-proline. Nature 176:501 (1955).
17. P.R. Schimmel and P.J. Flory. Conformational energy and configurational statitics of poly-L-proline. Proc. Nat. Acad Sci. USA 58:52 (1967).

18. I.Z. Steinberg, W.F. Harrington, A. Berger, M. Sela, and E. Katchalski. The configurational changes of poly-L-proline in solution. J. Am. Chem. Soc. 82:5263 (1960).
19. L. Stryer, and R.P. Haugland. Energy transfer: A spectroscopic ruler. Proc. Nat. Acad. Sci. USA 58:719 (1967).

INDEX

MIX
Papier aus verantwortungsvollen Quellen
Paper from responsible sources
FSC® C105338

If you have any concerns about our products,
you can contact us on
ProductSafety@springernature.com

In case Publisher is established outside the EU,
the EU authorized representative is:
Springer Nature Customer Service Center GmbH
Europaplatz 3, 69115 Heidelberg, Germany

Printed by Libri Plureos GmbH
in Hamburg, Germany